"I didn't realize how much I needed this book till I got it. I know it will help me in the identification of the wildflowers I photograph."

- Leonard Lee Rue III, PhD - Internationally known wildlife photographer with pics in several encyclopedias; author of 29 outdoor books, including *Deer of the World.*

"Your snappy definitions will be memorable for students (and others) who appreciate simple language."

- Edward G. Voss, PhD, Professor & Curator Emeritus, University of Michigan Herbarium, author of 3 volume *Michigan Flora.*

"Those that become 'experts' in a discipline eventually realize, at some point, that they really did it largely on their own with occasional boosts over the hard parts from other sources. Barb Short's instincts are right on, as it makes good educational sense to put an easy-to-read glossary up front into the hands of learners in Book 1. They will learn much that is useful by the time Book 2 appears."

-Douglas Valek, PhD, Professor Emeritus, Central Michigan University Botany Dept.

"The text was accurate and fun to read. I enjoyed the trivia and little known facts that can be presented to groups."

- Bruce Beerbower, Director of Visitor Services/Naturalist, Catawba Science Center, Hickory NC.

"... essential for properly identifying unfamiliar plants while in the field."

- Darrell Reider, Texas rancher/amateur botanist.

"Botany for Bloomin¹ Idiots should prove indispensable for all who wander through the botanical world, whether they are interested in identifying the wildflowers they find along the way or are primarily backyard gardeners. Learning the language is vital to deciphering field guides, botanical manuals, and even the standard gardening books. Author Short offers here the vocabulary needed to talk the talk and, more important, to understand it.

"The extensive glossary covers most of the terms that will be found in botanical manuals, which use a language all their own. Moreover, the section on the derivation of terms and scientific plant names makes the conversation that much more interesting. Finally, a chapter on the botanical and agricultural time line, from prehistory to the present, offers unparalleled insight into the chronology of plant development. The latter is a feature we have not seen previously and find enormously fascinating.

"Short¹s book occupies a central spot on our bookcase of botanical guides and manuals. We will use it often and feel sure others will find it equally useful and interesting."

-John & Gloria Tveten, Naturalists, writers, photographers, authors of *Wildflowers of Houston and Southeast Texas*

BOTANY
FOR
BLOOMIN' IDIOTS

BOOK I: TALKIN' THE TALK
GLOSSARY AND NAMES

BARB SHORT

Published by Fence Row Publishing
Marlette, Michigan

Botany for Bloomin' Idiots
Book I: Talkin' the Talk, Glossary and Names

By Barb Short

Published by:

Fence Row Publishing
Marlette MI 48453 USA

Copyright © 2007 by Barbara J. Short
First Printing 2007
Printed in the United States of America by Color House Graphics, Grand Rapids MI 49508

Library of Congress Cataloging-in-Publication Data
Short, Barbara J.
    Botany for Bloomin' Idiots-Book 1-Talkin' the Talk: Glossary, Timeline and Names
    by Barb Short - 1st Edition, 192 p.; illus, First Printing 2007
    Includes bibliographical references and table of contents
    LCCN 2007906749
    ISBN 978-0-9762714-3-7   Soft cover $29.95
1. Botany - Dictionaries   I. Title
2. Botany - Terminology
3. Wildflowers—Reference—Handbooks, Manuals, etc
4. Gardening—Reference—Handbooks, Manuals, etc
5. Humor—Gardens

1 2 3 4 5 6 7 8 9 10

comedy - *A form of dramatic literature designed to amuse and often to correct or instruct through ridicule. To achieve its effects, it exposes incongruity, absurdity, and foolishness, and its treatment of characters frequently has elements of exaggeration and caricature. –*

Encyclopedia Americana.

Talkin' Acknowledgements

Thanks to all those who contributed to this book:

Encouragement from authors: Lennie Lee Rue III, Walt Hoagman

For computer help: Beth McQueen, Nick Kitchen, Jerry Buda, Brook Atkinson, Mark Sheler

Proof reading and suggestions: members of the online Nature Writer Workshop, especially Lee Basner, Dee Walmsley, and Marge Hermans; Joyce Holsinger, Wayne Porter, Darrel Reider, Bruce and Becky Beerbower, Marilyn Smith, and Lissa Porterfield.

Accuracy and knowledge: Ed Voss, PhD, University of Michigan; Doug Valek, PhD, Central Michigan University; and the authors of all the materials in the bibliography.

Art advice: Trina Brown. Art instruction: Brian David Smith

All the friends and family who were always there to sustain me when it seemed this would never end, with special thanks to Phyllis King, Clare Bussjaeger, Jane Wright and Sophia Denton.

Any mistakes are my own.

Table of Contents

Didja hear about the elf who got tangled in the bellflower and tolled himself off?

EVERY CHILD SHOULD KNOW A HILL,
AND THE CLEAN JOYOF RUNNING DOWN ITS LONG SLOPE
WITH THE WIND IN HIS HAIR.
—EDNA CASSLER JOLL

These I've loved since I was little:
Wood to build with or to whittle...
Back country roads and cawing crows,
Stone wall with stile going over,
Daisy, Queen Anne's Lace and Clover...
Friendly dogs and friendly people
—Elizabeth Ellen Long

Some keep the Sabbath going to church,
I keep it staying at Home –
With a Bobolink for a chorister
And an Orchard for a dome.
— Emily Dickinson

GARDENING IS THE ART THAT USES FLOWERS AND PLANTS AS PAINT,
AND THE SOIL AND SKY AS CANVAS.
—ELIZABETH MURRAY

Anything will give up its secrets if you love it enough. Not only have I found that when I talk to the little flower or to the little peanut they will give up their secrets, but I have found that when I silently commune with people they give up their secrets also – if you love them enough.
—George Washington Carver

Nature will bear
the closest inspection.
She invites us
to lay our eye level
with her smallest leaf
and take an insect view
of its plain.
—Henry David Thoreau

All through the long winter, I dream of my garden.
On the first day of spring, I dig my fingers into the soft earth. I can feel its energy,
and my spirits soar.
—Helen Hayes

PART I

GLOSSARY INTRODUCTION

Botany students and wildflowerers tell me what scares them most is not trying to find the flowers, but talking to knowledgeable people afterward. You can overcome that fear by learning to talk the talk.

Think of this book as a work in progress. Use the wide margins to make notes and draw sketches. The book was started to help me remember. When people started asking me for copies, it was clear there would be a market. We can all work to add to it for our own use.

This book is planned to make botany fun. Enjoy sharing your experiences with other people. It isn't fun if you look stupid.

Any subculture group likes to be exclusive, and often outsiders are made to suffer until they become members of that group. Part of this insider identity is speaking a different language. Botanists are as guilty of this as any other subculture.

A glossary has each entry made up of two parts: pronunciation and definition. If the word is pronounced like common speech, pronunciation is not indicated. Some have sketches when that will clarify the definition more than words can.

PRONUNCIATION

Subcultures often invent their own pronunciations and meanings of common words, and everyone in the group understands those, but outsiders have no idea what they are talking about. This is true of botany as well as in law enforcement or street gangs or social work.

The purpose of speech is to communicate; if you use these pronunciations, the people you talk with will know what you mean. If they prefer to use speech as a weapon, making you feel inferior, they are not interested in sharing their knowledge with you. You can find someone more cooperative.

The important point is that if you use these pronunciations, the person you speak with will know what you are talking about. Most who use a word differently than you pronounced it, will use it in another way soon after, pronouncing it the 'correct' way. If they need to feel superior using a different pronunciation in an unpleasant way, that is their problem, not yours. Play their game, and use the word as they pronounced it. You may learn something in spite of them.

You will also find words pronounced differently in the North and the South, the East and the West. For example, you may find a plant identified as yarrow or Achillea (Ah kill'ee ah) in the North, while the same genus is called milfoil or Ah kill lee'ah in the South. All four terms are 'correct', in that they are understood in the area where they are common.

Don't panic or get embarrassed because you pronounce something in a different way. That person looking down their nose at you could be just as 'wrong' in a different group.

The pronunciations given after the technical words in the following list were chosen to follow pronunciation in Webster's Collegiate Dictionary. Instead of using all those little squiggles over the vowels that must be looked up each time, the pronunciations consist of words and sounds like printed American speech. **Don't try to break the words into syllables from these.** They are written the way you might speak them, not to write a technical paper.

Botanists and other subcultures will argue that a dictionary is only a history book of language, that it has not caught up with what is actually being spoken. Be that as it may, the pronunciations given are the commonly accepted ones, and if they are spoken differently in your group, switch to their pronunciation.

DEFINITIONS

Botanists don't always agree on definitions, especially between those botanists in the ivory tower of academia and those who work in horticulture. For example, HORTUS THIRD, the ultimate source of information for nurseries and landscape architects, describes a cyme: 'A determinate inflorescence, usually broad and more or less flat-topped, the central or terminal flower opening first.' In FUNDAMENTALS OF PLANT SYSTEMATICS, a textbook for scientific botanists, we find under 'unbranched inflorescences with pedicillate flowers', a simple cyme is described: 'A determinate, dichotomous inflorescence with the pedicels of equal length.' It is obvious that if the pedicels (stalks) are of equal length, the flower is not going to be flat-topped, but rounded. Therefore, the definition in this book takes the undisputed information from several sources consulted and defines cyme: 'Unbranched determinate flower clusters with stalks on the individual flowers.'

SKETCHES

Botanical sketches are by the author and included for those things that are nearly impossible to express in words. For example, the tips, bases and margins of leaves are illustrated. A 'typical' flower is printed in gray, with the part under discussion outlined in black.

None of the sketches is meant to be correct for anatomy of any particular species, just a generic example. If you are looking for great art, go to the Smithsonian. Hopefully these examples will help you understand and give you courage to do sketches of your own.

Wide margins supply room for your sketches, field notes and library additions. Most of us were raised not to write in books. Books become friendlier after we leave our marks on them. Pieces of paper float behind us like autumn leaves, but notes in our favorite books stay with us forever.

SOURCES

Botany, flora, systematics, wildflower and horticulture books (see Bibliography) and the internet (see Internet Sources) were researched to find which words might confuse you.

A

a- (ah; ay) – Combining form meaning without or not.

abaxial (ab axe' ee al) – The outside (bottom) or dorsal surface of a part. See dorsal.

aberrant (ab' er rant) – Departing from normal.

abbreviated (ah breve' ee ate ed) - 1) Shortened, as when one part is not as long as normal.
2) The type of tops worn by popular female students.

abortive (ah bort' ive) – Not fully developed, defective. The abortive flower will drop off as in violets, *Viola*. Abortive fruits will have no seeds, are called mummified.

abrupt - 1) Ending suddenly, not tapering, squared off.
2) The attitude of some botanists.

acaulescent (ak' aw les' cent) - Looking as though it has no stem, as when the leaves and flowers grow directly out of the ground.

accessory (ak sess' or ee) - Having anything included which is not really necessary but contributes to the complete picture. An additional or supporting part.

accessory buds - Extra leaf or flower buds at or near nodes rather than where the leaf joins the stem or at twig tip. There are two types of accessory buds. The buds at the sides of another bud are called collateral buds, and the ones above the bud are named superposed buds.

accessory fruit - A fruit that grows from parts of the flower other than the usual female part (pistil). For example, an accessory fruit may grow from the base of the flower (receptacle) as in the strawberry, *Fragaria*, with the seeds (achenes) embedded on the outside of the flesh instead of hidden inside.

accessory organs - Flowers as well as animals have organs. The organs most commonly referred to are sexual organs, female (pistils) or male (stamens).
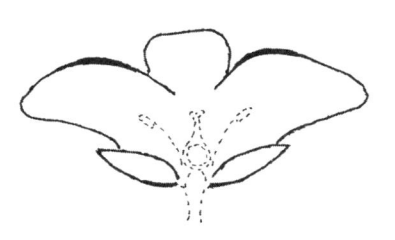
Accessory organs are parts of a flower that are <u>not</u> directly connected with breeding, such as petals and sepals. Synonyms: floral envelope, perianth.

acerose (ass' er ose) - Sharp, needle-like.

achene (ah keen')- A small one-seeded dry fruit with a tight, thin outer covering that does not split open at maturity (indehiscent). Often has a tail or hook or feather, such as a dandelion, *Taraxacum*, or beggar's ticks, *Bidens*; or may be located on the outside of a fleshy fruit like a strawberry, *Fragaria*. It is distinguished from a nutlet by having a thinner wall.

acicular (ay sick' you lar) - Needle-shaped foliage. In cross section, it may be either round or grooved.

acrid (ack 'rid) - Harsh and bitter in taste.

actinomorphic (ak' tin oh mor' fik)- A flower that is shaped in such a way that if you are looking into the face from any direction, the opposite sides are a mirror image. A complicated word for the simplest kind of flower. Synonym: symmetrical

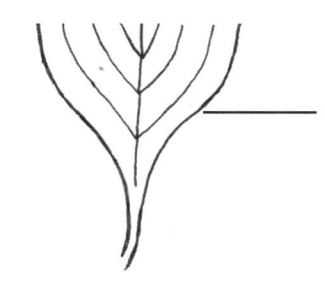

acuminate (ah cue' min ate)- Referring to the base or tip of a sharply pointed leaf or other structure, with the edges (margins) usually having a weak S curve.

acute (ah cute')- Referring to the base or tip of a leaf or other structure of less than a right angle, with the edges straighter than acuminate.

adenophorous – with various colored bumps or hollows that produce stuff (glands).

adhere (ad here') - To stick tightly but not be fused to, usually of dissimilar parts, as burs to your socks. Compare to adnate.

adherent (ad here' ent)- Clinging together of different kinds of parts, as men and women joining hands in a square dance or people of different tribes joining the intertribal dance at an American Indian Powwow. Adherent male and female parts are found in the mallows, Malvaceae, and others. See coherent.

adnate (ad' nate)- Having different organs of a flower grown together, such as a male part (stamen) attached to a petal. See connate.

adpressed (ad press'd) - 1) Parts flattened down to the surface but not joined, usually hairs on a stem or leaf. 2) Lying close and flat against, as a bud to a limb. Synonym: appressed. Antonym: divergent.

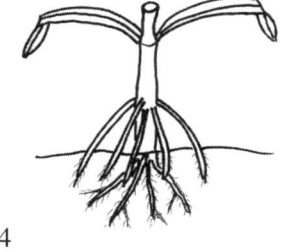

adventitious (ad' ven tish' ee us) – Adventurous, not where they are expected. 1) Said of roots poking out from the stem or leaf, as they do from the sides of stalks of corn, *Zea mays,* not those starting underground. These are prop roots and tips enter the soil. Compare with aerial roots.
2) In reference to buds growing randomly on the stem.
3) Sometimes refers to plants that have been introduced to an area, but are not truly naturalized.

aerial (air' ee al) - 1) Said of anything above the ground, as in describing the upper plant to differentiate from the roots.
2) What you need to make your TV work on overnight field trips.

aerial roots (air' ee al)- Those roots growing from the sides of the stem instead of underground, usually having the ability to penetrate the bark of the host on which a vine is climbing, as found on poison ivy, *Toxicodendron radicans*, and Virginia creeper, *Parthenocissus*.

aestivate (ess' tiv ate) – To hibernate in the summer instead of winter.

agglomerate, aggregate, clustered, crowded, conglomerate - Terms used more or less interchangeably, showing arrangement of leaves or other structures very close together, usually overlapping each other.

aggregate fruit (ag' rah gate) - Many small fleshy fruits (drupelets) with hard centers (endocarps), together in a cluster, like raspberry, *Rubus*.

agriculture – The practice of raising huge amounts of food for the masses, and the study of ways to improve the output.

air layering – A way to propagate plants that have lost their lower leaves and gotten leggy. Slit a node of the lower portion lengthwise, wrap damp sphagnum moss around it, and cover with plastic until adventitious roots develop. Cut and plant the shorter, leafier version. You feel so good about this that you cut the stem back to where leaves should be, and hope the roots will send you a healthy new top. Then you remember the cut-off stem might be a **cutting** so you stick that in some dirt.

alate (al'ate)- having parts that look like wings.

albino - A flower or other part that lacks normal color, is white.

alien (ay'lee en)- From another country, brought in on purpose or by accident. This becomes an issue when a species becomes successfully established, particularly if it displaces native plants as purple loosestrife, *Lythrum salicaria*, has done. Synonyms: exotic, foreign.

allelopathy (al' eel op' path ee) – The ability of plants to squirt chemicals into the soil that inhibit seeds from other plants to germinate, allowing the plant to compete successfully for water and nutrients.

allopatric (al'oh pat' rick) – Two or more species with separate and mutually exclusive geographic ranges.

alternate (all' turn nat)- 1) Arranged with one leaf or structure on each node, with every other one on the same side. Progressing without being opposite each other, for example, leaves along the sides of a stem like footsteps, left, right, left. 2) With other organs such as stamens and petals, alternate stamens are located between the petals instead of in front of them: stamen, petal, stamen. Alternate sepals are also between the petals: petal, sepal, petal. Compare with opposite, whorl.
(flower back shown)

alveolus (al vee ole' us) – A pit, pore or other small depression.

ament (ah' ment) – A spike (infloresence), usually all one sex and without petals; a catkin.

amorphous (ah morf' us) - Shapeless, the form not regular or predictable.

amphibious (am fib' ee us) - Capable of living either on land or in the water, like for-get-me-not, *Mysotis*.

amplexicaul (am plecks' ih call) - Clasping the stem, as the base of some leaves wraps completely around the stem.

anaerobic (an air oh' bick) – Describes organisms that survive without oxygen. Generally found in reference to compost heaps and students with sinus trouble.

anatomy (an at' om me) – The study of internal structure of something, like a plant.

androecium (an dree' see um)- All the male parts of a flower; the stamens. A guy thing, Andy is a guy.

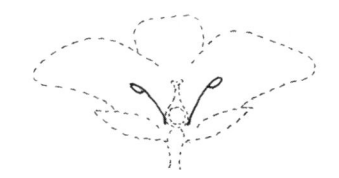

androgynous (an droj'in us**)** – Having both male (staminate) and female (pistillate) flowers in the same clusters.

anemophilous (an' em ahf' ill us) – Pollinated by wind rather than insects.

angiosperms (an' gee oh sperms) - Flowering plants having their seeds inside a closed organ (ovary). This includes most trees and grasses, as well as the plants that are commonly thought of as flowers and weeds (forbes). It does not include pines (conifers), ferns, mosses and fungi.

annual (an'you ul) - A plant that would have grown, bloomed, set seed and died in one season, if it had lived after you pulled it up and took it home. **Winter annual** - Starts to grow in autumn; blooms, fruits and dies the following season.

annular (an' you lar) - Ring-like, a solid in the shape of a ring, as growth rings on a tree.

annomalous (an nom' ah lus) – Not as expected, abnormal.

anterior (an tee' ree or)- Up front and personal, in front of the stem, midrib, pistil or other organ (axis). On a lipped flower, the bottom lip is **anterior**, the top lip is **posterior**. The anterior lip is sometimes called **inferior** or **exterior**.

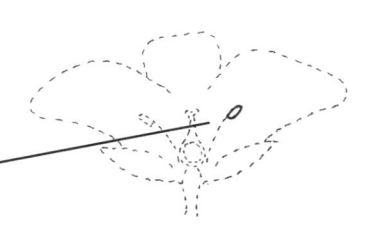

anther (an' thur)- The organ at the top of the (male) stamen, the structure that produces the pollen.

anthesis (an thee' sis) – Opening of the outer organs of the flower when the pistil and stamens are maturing. Synonym: blossoming.

anthropology (ann' throw pall' oh gee) – The study of people, modern or ancient, that searches for the origins and culture of the group.

antrorse (ant' roar s) – Pointing forward or up. Antonym: retrorse.

aperturate (ap purr' toor ate) – With one or more openings.

apex (ay' pex)- Tip, as of a leaf or petal. The part farthest from the attachment point, even if the top is hanging down. This idea may be confusing, !but makes sense when you consider that much of the technical part of botanical work is done in the herbarium with pressed specimens which may not be arranged the way the plant appeared in the field. Synonym: distal end. See appendix: flowers.

apical (ay' pick al) - At the tip, as a point at the tip of a petal or leaf, or a leaf at the tip of a branch.

apical bud - The bud that forms on the very tip of a branch, with no mark showing where a previous leaf had been attached (leaf scar) beside it. Synonym: terminal bud.

apiculate (ay pick' you lat)- Referring to a sharp (acute) leaf tip with a short, sharp, flexible point added.

appendage (ah pend' age)- A part fastened on a bigger structure, usually sticking out or drooping, !suggesting an arm or finger.

appressed (ah press'd)- 1) Flattened down to the surface, usually said of hairs on a stem or leaf.
2) Lying close and flat against, as a bud to a limb. Synonym: adpressed. Antonym: divergent.

arborescent (ar' bore ess' cent) - Treelike in appearance and size.

arboretum (are' bore ree' tum) - A botanical garden of trees and shrubs.

arborist (are' bore ist) – Certified position for someone who studied how to correctly treat healthy trees; and repair damaged or diseased trees.

archeology (ark' ee ol' oh gee) – The study of prehistoric people and their culture, including finding seeds of plants, gathering tools, etc.

arcuate – (ark'you ate) – Arching, bowed.

areola; areole (ah ree' oh la; are'ee ole) – 1) A squarish space between leaf veins, seen most easily on the back of the leaf.
2) The bump on a cactus that grows spines.

aril (air' ill) - A baglike outgrowth partly or completely covering a seed, as in Chinese lantern, *Physalis*; loved by crossword puzzle constructors.

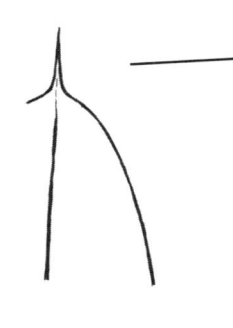

aristate (ah wrist' tate)- 1) Bearing a stiff bristle, or narrowing to a long sharp point, as at the tip of a leaf where the rib extends beyond the blade.
2) Prickly, as when many bristles are present on an organ.

armed, armor, armature - Refers to a plant that protects itself with thorns or other prickles. See appendix – armor.

aromatic (air' oh mat' ik) - Good smelling, especially crushed leaves.

arrangement - A non-technical term referring to the way things are put together, as a flower cluster (inflorescence) may be described by the placement of the flowers, or leaves as opposite or alternate.

ascending (ah send' ing)- 1) Rising upward at an angle, but not straight up, like your back when you have been leaning over too long to identify a new flower. When the spirit is willing, but the flesh—or stem—is weak.
2) Precisely, any line angling upward at a scale of 16-45 degrees away from the perpendicular. If wider than that, it is called **inclined.**

asexual (ay secks' you al) – Refers to vegetative methods of plants reproducing from means other than pollination. In nature, this includes stolons, runners, tubers, bulbs, etc. People have invented air layering, tissue culture, grafting, all usually practiced by those who end up on dates with obFred and obMargaret.

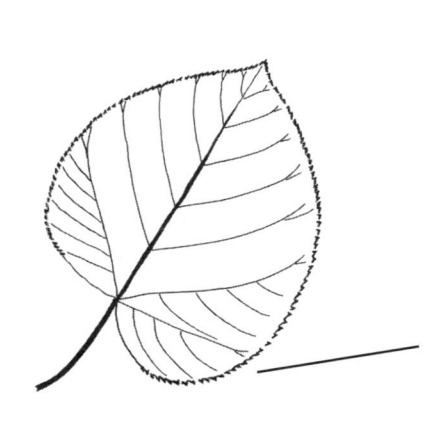

asymmetric (ay sim met' rick)- 1) Not a mirror image, for example, when one side of the base of a leaf has a pot belly (oblique), as in basswood, *Tilia americana*.

attenuate (ah ten' you ate)- A long gradual taper, as at the base or tip of a leaf or flower parts, drawn out into a long point, with a grass leaf being extreme. See acute.

atypical (ay tip ick al) - Not normal, uncommon.

auricle (or'ick ul) – With a lobe or appendage shaped like an ear.

author - The botanist who has named and described a plant in exact procedure, that name being accepted as the correct name for that species. The author's full name or abbreviation follows the scientific name (binomial epithet), as columbine, *Aquilegia canadensis L.*, with L. referring to the author Carl Linnaeus, who named many plants.

autumnal (aw tum' nal) - Appearing in the fall season.

awl-shaped (all) - Tapering from a wider base to a point, a slender 3-dimensional cone.

awn (to rhyme with lawn) – A specialized prickle found on the fruit (grains) of grass. See prickle; appendix – armor.

axil (aks' ill)- The angle formed by two lines (axes), as a leaf stalk off the main stem, like an arm off the body. See axis.

axile (acks' eel) - Belonging to, or found in, the axil.

axillary (aks' ill air ee)- In the plant's upside-down armpit, as axillary blossoms grow between the main stem and a branch, or between a branch and a leaf, instead of at the tip. Synonym: interfoliar.
Antonym: terminal.

axillary bud - A bud that forms on the side of a stem at the base of the
leaf stalk (petiole), found over the leaf scar in winter. Synonym: lateral bud.

axis, plural **axes** (axe' iss, axe' ease)- The main line of growth in a plant or organ, as the stem, from which the other parts such as the leaves and flowers grow.

B

b and b – A marketing term, ball and burlap. Small tree seedlings can be sold bare-root, but larger ones need to have the roots protected. Saplings have the roots (normally about as long as the tree is tall) pruned to a manageable size, then wrapped with coarse cloth (burlap) around moist soil to be transported. Burlap should be removed when the tree is planted. See bare root, sapling, prune.

Backyard Wildlife Habitat – Certification available for homeowners who provide food, water and shelter for wildlife.

balsamiferous (ball sam if 'er us) – Sticky and smells like pine pitch.

banded - With crosswise stripes of one color on another color. If the stripes are lengthwise, they are called (you got it!) striped.

banner - The high flying part of a flower of the legume family, such as pea blossom, *Lathyrus*. Synonym: standard.

bar – 1) A crosswise band of a different color.
2) A colorful watering hole.

barbate (barb' ate) – Bearded, with tufts of long weak hairs.

barbed - With short, hard bristles that are hooked. See prickle.

bare-root – A method of marketing trees, shrubs and some perennials, removing the soil from the root ball to cut down on shipping costs.

bark – The outer layer (epidermis) of a stem.

barren (bare' wren) – Not fertile, incapable of reproducing.

basal (base' al)- At the bottom.
1) In biennial and perennial herbaceous plants, the first year a rosette of leaves often grows low to the ground, called basal leaves.
2) May refer to the lower leaves on a stem. Antonym: aerial.

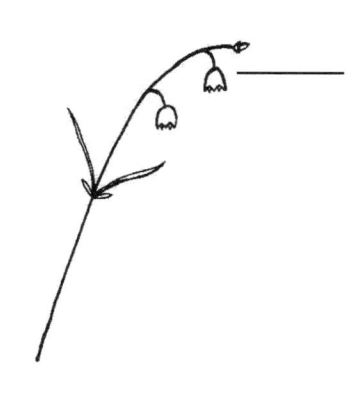

base - The bottom of any part, closest to the attachment point.
If the flower droops down, the flower base is located at the top.

bastard - A plant that has crossed with another species (hybridized) and is almost impossible to classify.

beaked - With an outgrowth like a bird's beak, as on a stigma or seed.

beard - A line of soft prickles, usually on a petal, especially if the hairs are a different color and fluffy, as some wild flags, *Iris spp*. See prickle.

bed – Ground that has been prepared to grow plants.

bedding plants – Mostly annual garden flowers for masses of color for the lasses who passes.

bell - Campanulate, shaped like a bell.

beneficial insects – Bugs like preying mantis, ladybugs and lacewing flies that eat bad bugs.

berry - A botanical berry is not what we usually think of as a berry. It is a fleshy fruit with a juicy (succulent) wall (pericarp), holding the seeds inside, like a grape, *Vitis*, or a banana, *Musa*.

biconvex (by con' vex) - A circular solid form that is rounded out on the two large outer sides, like an aspirin tablet. Synonym: double convex.

biennial (by enn' ee al)- A plant that builds up its 'muscles' one year, to be in shape for blossoming and seed production the second year. A biennial grows leaves and maybe stems the first year; blooms, fruits and dies the second year.

bifid (by' fid) - Cleft into two parts, like a leaf that grows split down the middle vein so it looks almost like two leaves.

biflorous (by floor' us) – Flowers twice a year, like spring and autumn.

bifoliate (by foal' ee ate)- With two leaflets growing from the same point at the top of a stalk (petiole or petiolule) of a compound leaf.

bifurcate (by' fur kate)- Forked into a Y-shape, like stems or branches.

bilabiate (by lay' bee it) - With two lips, like a snapdragon, *Antirrhinum*, a bilaterally symetrical flower.

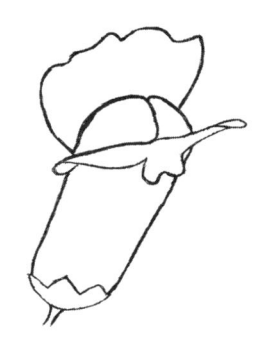

bilaterally symmetrical (bye latt' er alley sim met' rick al) – With a mirror image in only one direction. See bilabiate. Synonym: zygomorphic.

biological pest control – Using natural living enemies (like lady bugs) of bad bugs to protect your garden. See IPM.

binomial (by nome' ee al) - A two-part scientific name under the system devised by Carl Linneaus in the 1700's. Earlier, scientific names sometimes extended for a full sentence. The first part of a binomial, genus, is capitalized. The second part, species, is usually not capitalized. In painted trillium, *Trillium undulatum*, Trillium is the genus and undulatum is the species. Linnaeus wanted the binomial to describe the plant: something like Altissima (tall) phewii (stinky). Too often the botanist (author) decides to name it after himself and his mother: Smithianthus jerriatrica, so students call it "that tall stinky plant."

biological bribe - Anything offered by a plant to induce another organism to help. For example, food on a fruit encourages birds to carry the seeds for wide distribution; nectar deep inside a flower stimulates bees to carry pollen and promotes fertilization.

bisexual (by seks' you al) - Including both male parts (stamens) and female parts (pistils) in a flower. Synonyms: perfect, monoclinous.

bladdery (blad' er ee) –Like a small balloon.

blade - The true leaf, not counting the stalk (petiole).

blanch – 1) To turn pale from some lack, such as a lack of light when you threw your sleeping bag over a plant, and left it there all week.
2) What you do when the instructor announces a quiz and you lack knowledge.

bleeding – 1) The loss of sap from a wound.
2) What you do a lot of when classifying *Rubus* and *Amelanchier*.

bloom - 1) An easily rubbed off powder on a leaf or fruit, as on a blueberry, *Vaccinium*.
2) A non-technical word for a flower.
3) A genteel swear word, as, "I can't identify the blooming thing."

blotched - With a color pattern of broad, irregular spots.

BMP (bee em pea) – Best Management Practices. An agricultural program of improving growing practices. See IPM, crop rotation, organic gardening/farming, companion planting, monoculture.

bog – 1) A peatland showing many transition zones from mature forest or plain to open water, usually with a floating mat.
2) Commonly but incorrectly used for any soggy ground.

bole (bowl) – Stem of a tree, the tree trunk.

bolt – 1) Said of plants like lettuce, *Latuca*, when you want them to keep producing salad leaves, but they send up a stalk (scape) and get bitter.
2) What you should do when the farm dog heads toward you.

bonsai (bon' sigh) – 1) A horticultural process used by S&M folks to produce miniature trees. The limbs and roots are tightly pruned, tiny amounts of minerals and water are administered. See depauperate.
2) What you should yell when you bolt from the farm dog and fall off a cliff, to convince others that is what you meant to do.

border – 1) In reference to a flower made up of one color with a broad edge of another color. See edged, picot.
2) The low growing flower plants at the edge of a bed.

botanical garden (boat tan' ick al) - A zoo for plants, which are placed in beds instead of cages.

botanist - A person who talks so long you can't stay awake, but talks so loud you can't sleep.

botany – The science that covers anything to do with plants. The following groups appear under the title:
1) Structural botany, broken down into two parts:
   A) Morphology – shapes and development; and
   B) Anatomy, internal structures.
2) Cytology – Study of cells.
3) Physiology – The life functions of the plant and its organs.
4) Systematic – Classifying plants into similar groups and naming them.
5) Distribution – Where plants appear on the Earth. Synonym: Geographical botany; range.
6) Paleontology – Looking into the past through study of fossils; narrowed to plants, the study is Paleobotany.
7) Ecology (oecology) – Plants in relation to each other and to their environment.
8) Applied Botany – Where the Ivory Tower meets the ground: ways that plants can be used to make money, i.e., agriculture, horticulture, medicine, forestry, food preservation, landscaping, whatever. Synonym: Economic Botany.

bottle garden – 1) A terrarium made from a large glass bottle.
2) Your excuse for buying magnums.

bottom - The botanical bottom of a flower is the part closest to the point of attachment, so the bottom of a flower that is hanging down may be the top of the flower as seen on the plant. See top, base.

brackish – Water that is rather salty, as found in estuaries.

bract (brakt)- 1) A modified leaf in the inflorescence, usually small, sometimes reduced to a papery scale. However, it may be huge and colored, as in the Christmas poinsettia, *Euphorbia*, or a hood (spathe) over a Jack-in-the-pulpit, *Arisaema*, or medium-size and looking like a petal, as in dogwood, *Cornus florida*.
2) A loose description of any small leaf-like projection.

bractlet (brakt' let) - A tiny leaflet, such as that found at the base of an individual flower stalk (pedicel) on a compound cluster of flowers (inflorescence), like Queen Anne's Lace, *Daucus carota*.

bramble (bram' bul) – See prickle.

branch – 1) A non-technical term referring to the side shoots of a tree trunk, flower stem or flower cluster.
2) How you order water with your drink at the bar.

brevi- (brev' ee) - Combining form meaning short, brief.

briar, brier, bristle (bris' tul) – See prickle.

Bryophyte (bry' oh fight) – A flowerless plant from the great group Bryophyta, that includes the mosses, liverworts and hornworts.

bud - A baby leaf or flower, with all the essential parts, but in an immature state (rudimentary). A bud may occasionally hold both leaf and flower (mixed bud). The bud may have scales around it (protected) or not (naked).

bud scale - A modified leaf or bract, seldom green, protecting a bud.

bugs - Hungry, crawly things that get into your herbarium specimens and turn them into sawdust (frass) the day before you were supposed to take them to class.

bulb - A storage organ made of the stem surrounded by modified leaves squeezed together into a ball, usually white inside, buried down with the roots, like an onion, *Allium cepa,* which is a concentric bulb with layers arranged like a target.

Another type of bulb is found in garlic, *Allium sativum.,* which some books call naked bulbs, and others call bulbils.
Cooks call them garlic cloves. See corm and fruit.

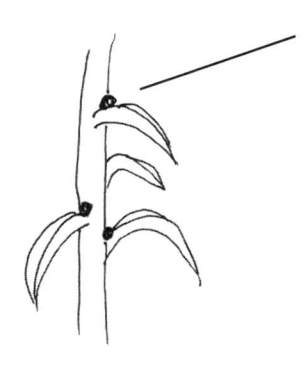

bulblet or **bulbil** (bulb' let, bulb' ill) - A bulb that was too little to be out alone and got lost, ending up in the leaf axil, as in Turk's cap lily, *Lilium superbum*, or some other weird place a bulb shouldn't be.

bundle scar – The mark left on a branch by the stalk (petiole) and vascular bundles (xylem and phloem) when a leaf falls off. See scarred.

bur – 1) A fruit having the bractlets (involucre) formed into a hollow vessel (accessory fruit) with the seeds inside and barbs outside, as in burdock, *Arctium major*. See prickle; see appendix – armor.
2) A common word when learning winter identification.

burl - 1) A swollen area on the trunk of a tree.
2) A method of preparing edible fruits and vegetables.

buttress (butt' tress) – Supporting wing on the tree trunk.

C

c.; ca. – Abbreviation for circa, meaning "around", as "Discovered ca. 1852" or "Lives c. 10 years."

caducous (kah duke' us) – falling off early, before maturity.

calcarate (kal' kar ate) - With tails or spurs, as a columbine flower, *Aquilegia*.

callus (kal' us) – A thickened layer, often hard.

calyx (kay' licks, kal' licks)- The whole group of leaflets (sepals) under a flower together, ring around the rosy, that cover the petals in a bud. Usually green and may be separate or appear glued together at the base (united). Synonym: outer perianth

cambium (kam' be um) The growth layer between the bark and woody part of a plant, that causes the stem (trunk) to become fatter, not taller. This must be in contact with the scion in grafting. See grafting, scion.

campanulate (come pan' you lat) – Bell-shaped, as some united petal corollas.

canaliculate (can al lick' you late) – with channels or grooves the long way (longitudinal).

candle - 1) Tender spring growth of conifers, Pinaceae.
2) What must be burned at both ends to pass a botany exam.

canescent (kah ness' ent)- As hairy as a canine. Gray with surface hairs.

canopy (can' oh pea) – The upper layer of foliage in a forest.

capillary (cap' ill air ee) - Extremely slender, hair-like part.

capillary action – The ability of water to climb when confined, like up the side of your glass of iced tea, or inside of small vessels like plant veins, or between particles of soil.

capitate (cap' ih tate) 1) With a single head-like organ, as a stigma shaped like a mushroom cap.
2) Referring to a cluster of flowers tightly crammed together (head).

capsule (kap' sool, often slurred to kap' sul) In flowering plants, a dry fruit that opens (dehisces) along two or more lines (sutures) to release two or more seeds. For illustration of several types, see dehiscence.

carinate (care' in ate) – Keeled with ridges running the long way (longitudinal).

carnivorous (car niv' er us) – Meat eating. Said of plants that live in soil that is so acid that the plant cannot absorb nutrients adequate for growth and reproduction. These plants trap insects (meat) to collect nitrogen.

carpel (carp' ul)- The female unit used in making up a pistil. A **simple pistil** has one carpel, while a **compound pistil** has two or more. You can often figure the number of carpels by counting the stigmas, styles, or bulges on the ovary.

cartilaginous (car' till aj' in us) – In reference to tough tissue that has no chlorophyll or veins, like cartilage.

catalog – Beautifully illustrated work of fiction that arrives at New Year.

catkin (cat' kin)- A compact, bent-over spike of small stalkless (sessile) flowers, usually all of the same sex (either male or female but not both; unisexual), and generally without petals. The flowers grow on a slender flexible stalk in close circular rows. Found in poplars, *Populus*, and birches, *Betula*.
Synonym: ament.

caudate (caw' date)- With a little tail like a puppy dog's, as on the tip of a leaf.

caulescent (caw less' cent)- With a sturdy stem above ground, often a leafless flower stalk.

cauline (caw' lean) - Referring to the stem, often about something on the stem, or in comparing the upper plant parts to the lower (basal) parts.

cells - See organs.

centrifugal (sen triff' oo gull)- Said of a flower cluster (inflorescence) that starts blooming in the middle and proceeds to the outer rim. See determinate.

centripetal (sen trip' eh tull)- Said of a flower cluster (inflorescence) that starts blooming on the outer rim with the center flowers blooming last. See indeterminate.

cernuous (sir' new us)- 1) Droopy, as capsules hanging down.
2) With the tip drooping.

chaff - 1) Papery scales or bracts, especially between the seeds of the flowers of the daisy family, *Compositae*, and around grains of the grass family, *Graminae*. Synonym: palea.
2) Used loosely for any part of a dried plant that breaks loose, as leaves from hay or in the herbarium drawer after the first year botany students have handled specimens. See seed.

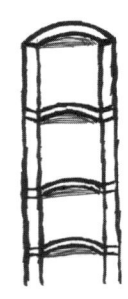

chambered pith - Having no solid core of pith, instead having walls crosswise of the stem, leaving hollow spaces between the walls. Compare to **diaphragmed pith** that has a solid core of pith in addition to the walls across.

chartaceous (car tay' shus)- With thin texture like charts, not green.

chasmogamous (kaz mog' a muss) – A normal flower that opens for pollination. See cleistogamous.

child sling - A device worn to carry children on field trips, from which baby boys compare how good their parents are at acceleration on hills.

chilling injury. Damage to certain tender plants, like cucumber and sweet potato, that happens from exposure to cold but above-freezing temperatures. See winter kill, die back.

chilling requirement. A cold period necessary to certain plants and plant parts to break dormancy or rest. The chilling requirement is measured by the required number of hours at 7 degrees centigrade or less. Apples, *Malus*, are especially sensitive to this in order to set fruit.

chlorophyll (klor' oh fill) – Green coloring matter of plants that produces food for the plant.

chlorosis (klor' row' sis) – An abnormal yellowing or blanching of leaves due to lack of chlorophyll. May be caused by inadequate light, insufficient iron, or other lacks.

chordate – See cordate.

cilium; pl. cilia (sill' ee um; sill' ee ah)- Hairs like eyelashes, as on the edge of a leaf or petal.

ciliate – With hairs along a margin or in a row.

ciliolate (sill' ee oh late) - With tiny hairs like eyelashes.

cinereous (sin near' ee us)- Gray in color, ashy, as though it had been !incinerated. Appearance due to covering of short hairs.

circinate (sir' sin nate)- Rolled from the top downward with the tip in the middle, like a snailshell or young fern.

circumboreal (sir' come bore' ee al) – With a range completely around the North Pole and somewhat southward.

circumscissile (sir' come siss ul) – In reference to one process of opening (dehiscense) a dry fruit. See dehiscense.

cirrhose (sir ose') – With a flexible tip at least 10 times as long as it is wide, and coiled. The tip may actually twine like a tendril.

clambering - A non-technical term meaning vinelike or climbing. See vine.

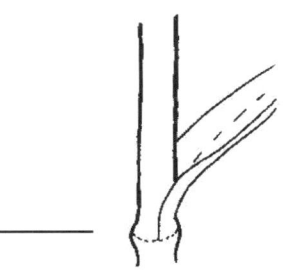

clasping - Refers to the position of the base of a leaf stalk (petiole) or lower edges of the leaf blade when it wraps around the stem, but isn't joined on the far side. See sheathe.

clavate (clay' vate)- Shaped like a baseball bat.

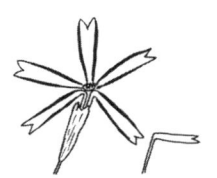

claw - A long narrow part with which the petal or sepal grabs hold of the base of the flower.

cleft - Where a petal or leaf margin or tip is divided about half way to the middle to form a deep notch. Like the outline of a cleft chin. See lobed.

cleistogamous (kl eyes tog' ah mus) - In reference to a specialized flower that never opens, forcing it to self pollinate, as in the hidden green summer flowers of violets, *Viola*.

climbing - Growing upward while holding on to a host or post by tendrils, petioles, adventitious roots, suckers, or the stem itself twining.

clouded - With colors unequally spaced like fluffy clouds.

cluster - 1) A non-technical term referring to a whole bunch of flowers growing together. See umbel, cyme, spike, corymb, panicle, raceme.
2) An aggregate fruit, with many ovaries, as a raspberry, *Rubus*.
3) A **bundle cluster**, as needles on *Pinus strobus*, white pine.

clustered, conglomerate, agglomerate, crowded, aggregate - Terms used interchangeably, showing arrangement of leaves or other structures very close together, usually overlapping each other.

coalescent (koh' ah less' cent) – Separate organs that grew together.

cochleate (cock' lee ate) - A solid that is coiled around like a snail shell, as in cockles and mussels, alive, alive, O.

coherent (ko here' ent) - Sticking together, usually of same kinds of parts such as anthers, like boys playing crack the whip. See adherent.

collar – In grasses, the base of the grass blade surrounds the stem. The point where the blade flares away from the stem is the collar.

collateral bud (koll at' er ul) – 1) An extra (accessory) bud placed above a bud on the side of a branch.
2) Your roommate who co-signs for your loan.

colonial – 1) In reference to a group of plants where all the members grow from one root system.
2) Loosely, any plants that grow in clumps.

column (koll' um) – The formation with the male (stamen) and female (pistil) parts joined together, at least at the base. Seen in orchids, Orchidaceae, and mallows, Malvaceae.

columnar (koll um' nar) - Referring to a plant that grows straight up with a stout main stem like a !column, with branches usually angled upward.

coma (koh' mah)- 1) A tuft of soft hairs as on a milkweed seed, *Asclepias*. A comet with a tail.
2) A tuft of leaves, as in pineapple fruit, *Aranus*.

comb – Prickles, soft or stiff, resembling a hair comb or cockscomb. See prickle.

commissure (com' ish your) – A joint or seam.

common name - The title that a plant is known by, often fanciful, and regularly different in parts of the country, as *Achillea millefolium* is yarrow in the north, milfoil in the south, and called by a word meaning 'squirrel tail' in some Native American languages. Common names often refer to uses, with 'wort' signifying a medicinal use, as motherwort, *Leonurus cardiaca,* or refer to history or folklore. Popular plants may have many names, with 23 being found for *Matricaria chamomilla*. Reference to an event, such as Mayflower for species blooming in that month, may be applied to an equally large number of species. Binomial epithets, recognized worldwide, attempt to overcome the problem.

comose (ko' mose) – See coma.

companion planting – The selection of different plants that enhance the growth of each other, provide barriers between monoculture, and retard the spread of pests and plant disease.

complete – In reference to flowers that have all of the inner and outer floral envelopes (perianths) plus the sex organs (stamens and pistils), all the common parts of a perfect flower.

composite (come pa' sit) - Made up of many individual parts, like the daisies, asters and sunflowers of Asteraceae (Compositae), are composed of many disk and ray florets in a head.

compound - Anything made up of two or more parts, as two carpels in a pistil make it a compound pistil.

compound leaf – A single leaf made up of two or more leaflets. See pinnate and palmate. Look at the base of the leaf stalks on woody plants; if there is a leaf bud at the bottom of each leaf stalk (petiole), they are simple leaves on a stem. If there is a leaf bud only at the base of the 'stem', you have a compound leaf, with leaflets on a stalk (rachis). Herbaceous plants are number two, and you have to try a little harder to tell the difference between simple and compound leaves.

compressed - Refers to a solid object flattened on the sides, looking like they had been ironed, like the leaf stalks (petioles) of quaking aspen, *Populus tremuloides*.

compulsiveness - A good thing for botanists to practice in moderation.

community – All the plants living together in a habitat.

concentric rings (kon sent' rick) – Patterned like a bull's-eye target.

conduplicate (con due' plea cat) - 1) Said of a leaf which is folded lengthwise along the center line (rib); V-shaped in cross section.
2) With the sides folded together so they look like the next to last step in making a paper airplane; often said of leaves or petals in a bud.
3) In a seed, stem or other organ, refers to a part with a groove running down one side, in which the cross section is more or less kidney shaped.

confluent (kon' flew ent) – Blending into one, running into each other.

congested – Crowded together.

conglomerate, agglomerate, crowded, aggregate, clustered - Terms used interchangeably, showing arrangement of leaves or other structures very close together, usually overlapping each other.

conical (kon' ih cull) - Refers to a solid shaped like an !ice cream cone.

connate (kon' ate)- 1) Joined at the base, as paired filaments of stamens.
2) United, as petals joined into a tube. Compare to adnate.

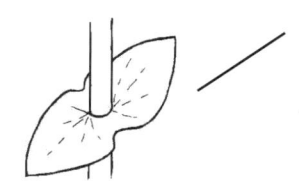

connate-perfoliate (kon' ate per foal' ee ate)- Opposite leaves joined at the base with the stem going through the middle like it was perforating or penetrating the joined leaves. Boneset, *Eupatorium perfoliatum*, is a good example.

connivent (kon nive' ent)- Getting their heads—or other parts—together, as if they were conniving up a practical joke. Coming close but not actually joined. Synonym: contiguous.

constricted (kon strik' ted) - Pinched in, as a pod between the seeds.

contiguous (kon tig' you us) - Touching or in contact, without actually joining. The lower 48 US states are contiguous, with the other two separated geographically.

continuous pith - Having the stem center filled with material, usually spongy, and white or green, that has no obvious chambers.

contorted (kon tort' ed)- All bent out of shape, as the petals in a bud.

contracted (kon track' ted) - Pinched in, as a pod between the seeds, like wearing a waist nipper corset. Synonym: constricted.

convolute (kon' vol lute)- Rolled up or twisted together lengthwise, especially as leaves or petals in bud. See revolute and involute.

copious (cope' ee us) – Extremely something, such as leaves or sap.

coppice (cop' iss) – 1) A small grove of young trees.
2) All the suckers growing from the stumps in a cut-over woodland. See suckers.

cordate or chordate (cord' ate)- Heart shaped, as in the base of a violet leaf, *Viola*.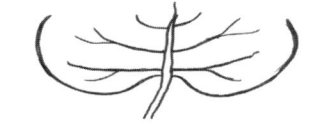

coriaceous (core ee ace' ee us)- With a texture like leather, as in liCORice sticks.

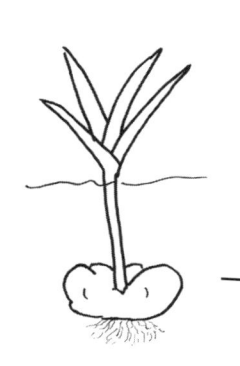 **corm** - An underground storage stem that is often mistaken for a bulb. A corm has a solid center while a bulb has layers like an onion or fleshy scales like garlic, both *Allium*. See bulb.

cormel (core mell') - A small corm that forms at the base of the mature corm.

corniculate (corn nick' you lat)- 1) Made of horny tissue.
2) With horns, as a milkweed flower (Asclepidaceae).

corolla (core roll' ah)- All the petals of a flower. May be separate (free) or joined together (united). The inner perianth.

corona (core own' ah)- 1) A crown on a flower, like the middle trumpet of a daffodil.
2) Crown-like growth on a petal or stamen.

coronate (core' on ate) - With a crown, as the united center petals of a daffodil, *Narcissus*. The whole daffodil flower is said to be coronate; while the center is a corona.

corrugate (core' rug ate) - With the surface having parallel ridges, like corrugated cardboard.

corymb (core' rim b) - A cluster of stalked (pedicillate) flowers, with the cluster having a flat or rounded top. The outer flowers open first (centripetal) and the stalk continues to grow (indeterminate). A simple corymb has a straight stalk (peduncle) dividing into alternately arranged flower stalks (pedicels).

costate (cost' ate) – With ribs, usually with ridges the long way.

cosmopolitan (koz' moh pol' it an) - Found all over the world.

cotyledon (cot' ill lee' don) - Seed leaf that appears when a plant germinates, and looks different from the leaves that will come later.

creeping - Running along the ground and sending out roots from the stem as it goes. Decumbent has the tip rising up; procumbent means it does not root. Synonyms: repent, prostrate.

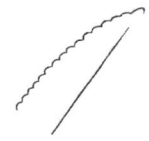

crenate (kren' ate)- Scalloped around the edges; with rounded teeth.

crenulate (kren' you late) - With small rounded teeth.

crest - 1) Fan-like growth.
2) A line of projections like a rooster's comb, as on a stigma.

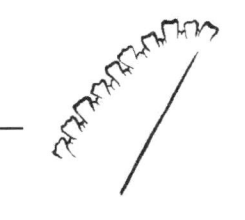

crispate (kris' pate)- With curled or ruffled edges, seen in the leaves of yellow dock, *Rumex crispus*. The progression of curliness from least to most is undulate, crispate, cirrate.

cristate (kris' state) - With a crest.

Cronquist, Arthur J. – The American who stood the world of botany on its ear in the late 20th century. Building on the work of Takhtajan, Bessey and Dahlgren, using modern tecniques like DNA and electron microscopes, he realigned the botanical families. Cronquist died in 1992; Takhtajan and others continue the work.

crop rotation – Planting different crops in a field each year, making it difficult for pests of one to survive in that field until the first crop is repeated. Each crop uses slightly different combinations of nutrients so the soil retains more balance.

crowded, aggregate, clustered, conglomerate, agglomerate - Terms used interchangeably, showing arrangement of leaves or other structures very close together, usually overlapping each other.

crown – The part of a plant where root and shoot meet.

crucifer (crew' siff er) - 1) Shaped like a cross; said of flowers with four petals or other parts arranged in the shape of a cross.
2) A member of Cruciferae, the mustard family, that have this type of flowers.

cruciform, cruciate (crew' see form, crew' she ate) - With four points, cross-shaped like a crucifix.

crustaceous (crust tay' shus) - With a hard, thin, brittle texture.

cryptogam (krip' toe gam) - A plant that reproduces by spores instead of seeds: ferns, mosses, algae, fungi. Though they may be puzzling, they are not to be confused with cryptograms, letter substitution puzzles.

culculate (kulk' you lat) - Hood-shaped, as the upper petals of milkweeds, Asclepidaceae.

culm (kulm) – A hollow or pith-filled slender stem like grass or sedge.

cultivar (kull' tiv are) – A plant that will breed true to form, and has been named to recognize that, often romantic names, like 'Heavenly Blue' morning glory.

cuneate (cue' knee ate) - Wedge-shaped, as a triangular leaf attached at one of the points. **Deltate** is attached at one flat side of the triangle.

cusp (cuss p) - A sharp, abrupt and often rigid point. See prickle.

cuticle (cute' ih cul) – A waxy layer on the surface of something.

cutting – A portion of the vegetation, such as leaf, shoot, root, stem or bud, that may propagate a plant without using seeds (asexually).

-cyclic (sigh' click) - Referring to the total of circles of different structures in a flower, commonly used with a number in front of it. For example, mullein, *Verbascum*, has 1) 5 sepals, 2) 5 lobes on the corolla, 3) 5 stamens and 4) an ovary, so it is 4-cyclic. However, it has 5 of most of the parts, so it is 5-meric. See -meric.

cylindric (sill in' drik) - With a long tubular shape.

cyme (sigh m)- Flower cluster (inflorescence) with stalks (pedicels) on the individual flowers, usually broad and rather flat. Middle flowers usually come into bloom first, which causes the stem to stop growing (determinate). Flower stalks (pedicels) do not branch so each holds only one flower.
    Helicoid cyme resembles a spring and has individual flower stalks branching all the same direction, such as to the left.

D

damping off – A condition often found when starting seeds or cuttings, fungus among us. The seedlings usually die.

dark dependent seed – A seed that must be in the dark to germinate, usually buried in soil. See light dependent seed.

DBH (dee bee aitch) – Diameter at Breast Height; the distance across a tree at about four feet up. This is used to calculate the healthy distance the next tree should be (one inch dbh = one foot away), and also helps to determine the board footage in a woodland.

dead-heading – Removal of droopy faded flowers to encourage more flowers to bloom. Be sure to remove the ovary and not just the petals. These old flowers have usually been pollinated. The job of a flower is to produce seeds. Once accomplished, the plant feels its work is done so it rests or dies. By preventing seeds, you encourage the plant to keep working.

decem- (dee sem') – Combining form meaning ten of something.

deciduous (dee sid' you us) - 1) Falling off, as petals or sepals after a flower blooms. Back of flower shown with sepals which may be deciduous, placed opposite the petals. 2) Describing trees that are not evergreens, and drop their leaves in the autumn.

declinate (deck' lin ate) – Curved downward.

decumbent (dee come' bent)- Said of branches lying down, but with the tips rising upward, like your head when you try to do sit-ups. See creeping.

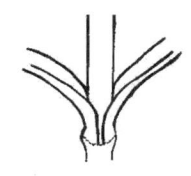

decurrent (dee cur'ent) – In reference to a part that extends down and along another part, like the bases of these leaves along the stem.

definite (deaf' fin it) – Said of a character that has been defined to narrow parameters, not flexible.

deflexed (dee flecks'd) - Bent sharply downward.

defoliate (dee foal' ee ate) - To cause loss of leaves.

dehiscent (dee hiss' ent)- The opening of a dry fruit, often a capsule, to release seeds by any of several methods.

Circumcissal type (sir come sis' al) flips its lid to expose the seeds,

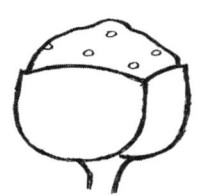

Poricidal type (pore ih side' al) spills seeds through little holes,

Septicidal type (sep' tee side al) splits wide open with seeds between papery separations (septa),

Locucidal type (lock' you side al) splits open to show seeds in spaces(locules), but doesn't have any separators.

deliquescent (dell ih kwes' ent)- 1) Turning to liquid. Instead of dropping its petals, the petals (or other parts) melt into a sticky glob, like spiderwort, *Tradescantia spp*, why it has the common name of snotweed.
2) Repeatedly branching into finer branches, like elms, *Ulmus*.

deltoid, deltate (dell' toid, dell' tate)- Triangular, like a river delta, attached on the broad side rather than a point, as a leaf like cottonwood, *Populus deltoides*.

dense: 1) Congested, as many flowers crowded together in a cluster (inflorescense).
2) How you feel when botany test papers are passed out.

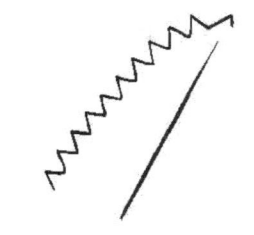

dentate – A margin with large teeth pointing outward, the teeth large enough for a dentist to work on.

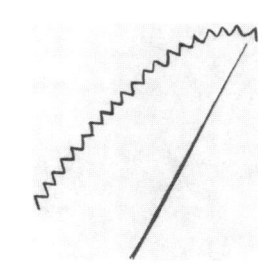

denticulate (den tick'you late) – With small toothlike projections, like a saw blade.

depauperate (dee paw' per ate) - Stunted or poorly developed, as starved as a pauper.

depressed - 1) Low, as if pushed down.
2) Angled downward. See descending.

depressed line - Groove. See conduplicate.

descending, depressed, reclined - Angling downward. **Depressed** is sharply angled downward at 166-180 degrees, like your arm hanging at your side. **Descending** is at an angle of 136-165 degrees, like your upper arm when you are relaxed and have your thumbs hooked in your belt. **Reclined** is bent at 106-135 degrees, like your arms when you are mad at botanists for using big words, and plant your fists on your waist.

dessicate (dess' ih cate)- Dry out, like a seed capsule.

determinate (dee term' in it)- Said of an inflorescence in which the central flower in a flat-topped cluster— or the flower farthest from the main stem in a pointed cluster — opens first and determines that the stalk won't grow longer. See indeterminate, terminal.

dewy – In reference to a surface covered with waxy platelets which appear to be dew drops.

diandrous (dye and' russ) - Having two stamens, di- meaning two and -androus meaning male.

diaphanous (dye af' an us) - With almost clear tissue.

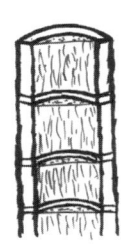

diaphragmed pith (dye' ah fram'd) - Having a solid or spongy core of pith with walls across at intervals, like grape vines, *Vitis*.

dichotomous (dye kot' om us)- Forking into a Y, then each of the tips forking into a Y again, return to Go, and repeat until tired. Synonym: bifurcate.

diclinous (dye kline' us) - Describes a plant with some flowers with male organs (stamens) and some flowers with female organs (pistils), no flowers with both male and female organs (perfect). Synonym: monoecious.
See appendix - breeding.

dicots (dye' cots)- The dicotyledonous plants, with two seed leaves (cotyledons), net (reticulate) veins in the leaves, and usually four or five petals on the flowers. See monocots.

didymous: (did' im us) – Twins, in pairs.

didynamous (did dye'nah muss) – With four stamens arranged in two pairs, one pair longer than the other pair.

die back – Branches that have stopped growing and surrendered, often to climatic changes of high or low temperatures, lack of rain, etc. They should be removed from plants. See winter kill.

diffuse (diff use') – Kind of floppy, loose, many branched.

digitate (dij' ah tate) – Like fingers, as a palmate leaf.

dilated (dye' lated) - 1) Enlarged, wide open, distended.
2) Flattened and broadened, like a fat filament.

dimorphic (die morf' ick)- Taking two different forms or outlines, though it is the same species, as in the fertile form called snakegrass, and the ferny sterile form called horsetail, both *Equisetum*.

dioecious (die eesh' us)- With male (staminate) flowers on male plants and female (pistillate) flowers on female plants. The oec- part comes from the same base as ecology (meaning house), with di- meaning two, therefore, the flowers come from two houses or plants. See monoecious and monoclinous. See appendix – breeding.

diploid (dip' lloyd) – With two sets of chromosomes. See haploid, polyploid.

disarticulating (dis' art tick' you late ing) – Describing an axil (stem, seed holder, etc) which falls apart at the joints at maturity.

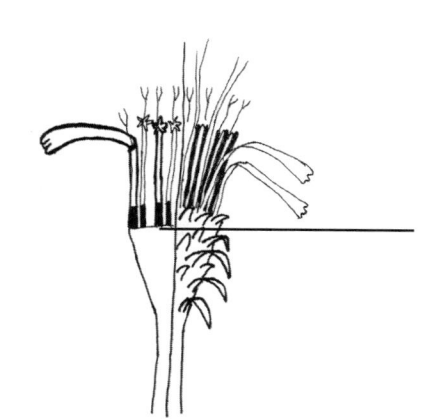

disc, disk - 1) Any circular, flattened organ.
2) Central group of florets in Asteraceae (Compositae).
3) Describing those individual tubular florets making up the central group of a daisy, aster or sunflower, called disc florets.
.

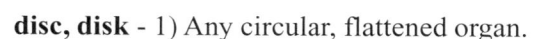

discoid (disk' oid) - 1) In reference to a flat circular solid, often with dents in either side like a breath mint (concave).
2) Referring to a floret of Asteraceae (Compositae) that is in the central disk of the head, not a strap or ray flower.
3) Shaped like a disk, said of female organs (stigmas) with several arms with disks at the tips.
4) Referring to an Asteraceae (Compositae) head without ray florets.

discoidal color - One large spot of color in the middle of another color, like an eye.

discrete (disk' krete) - Distinct, separate, said of parts such as petals and sepals that are not united. Synonym: Separate.

disjunct (dis junk' t) – Geographically separate from the main population, perhaps from seeds carried by a glacier or in a bird's belly and dropped in a receptive spot.

disk flower or floret – See disc.

disparate (dis pair' at) - Unequal, not similar.

dispersal systems (dis purse' al) - Methods of moving seeds away from the parent plant. This may include, but not limited to (I have always wanted to say that…) gizzards of birds, droppings of mammals, explosive plant mechanisms, rivers, wind, and pockets of curious kids.

dissected (diss eck' ted; saying dye' seck ted is a good way to make a botanist cry) –
1) Deeply cut into separate slender segments.
2) Said of a leaf that is deeply sectioned, but not to the midrib.
3) With various organs or tissues appearing to be cut apart.
4) Actually cut apart, as a specimen might be opened by a botanist.

distal (dis' tul) - Located far from the point of attachment, at or toward the tip (apex). Synonym: apical. Antonym: proximal.

disturbed – 1) In reference to areas that people have used, like farmland, old homesteads, etc. While searching for disturbed sites, such as anthropologists seeking Native American Indian sites, plants like staghorn sumac, *Rhus typhinia*, -- called indicator plants --in disturbed sites, can speed the location.

diurnal (die yearn' null)- 1) Opens only in the daytime, like a morning glory, *Convolvulus*.
2) Sometimes means something appears for less than a day (ephemeral).

divaricate (die vair' ih kate)- 1) Spreading far apart, extremely divergent.
2) Like an angler demonstrating the size of the fish that got away, which may also include prevaricate. See divergent.

divergent (die verge' ent) – 1) Going off in opposite directions, said of anthers that are attached oppositely on the filament, then droop down.
2) Spreading so the parts are separate, but not as far as divaricate, like your toes in thong sandals, said of buds pointing away from the twig. **Appressed** is cuddled up close, **divergent** is slightly angled, **divaricate** is strongly angled, but not quite a right angle, which would be called **horizontal**. Some books use divaricate and divergent as synonyms.

diversity (div verse' it ee) - 1) Refers to all of the kinds of plants, with about a quarter million species named and still living, in the world.
2) Describing large numbers of each level of classification, with 122 species of sedges, *Carex*, showing great diversity within the genus.
3) Referring to a difference in organs, as nine fruit types in species of roses, *Rosa*.

divided – 1) Botany - Separated three fourths or completely from the margin to the base or midrib in a leaf.
2) Horticulture – Hacked a root section into two or more parts to vegetatively reproduce several new plants.

Doctrine of Signatures – The belief that God marked plants with a clue to their use, e.g., boneset, *Eupatorium perfoliatum*, has leaves that are joined at the base so the plants will cause broken bones to heal.

dominant (dom' in ant) – In reference to the more active inherited trait (gene). See recessive.

dormant (door' mant) – 1) Refers to a seed that developed ways of protection so it doesn't sprout at the wrong time and die. Dormancy can make it hard to germinate seeds at home, unless you can break dormancy for that species, such as freezing, toasting, soaking, cracking, etc. Double dormancy is not uncommon, a period of cold, followed by moist warmth.
2) The period during cold weather when a plant has dropped it leaves and is resting, an ideal time to prune since the sap is not running and the wounds will not bleed. See resting period.

dorsal (door' sull)- 1) The back or lower side, as of a leaf.
2) Attached to the back or outer surface, just like the dorsal fin on a fish. Antonym: ventral.

dotted - Small spots of one color on another, but the spots are larger than what is called spotted. Polka dots. See punctate.

double - 1) Describing a flower with more than the usual number of petals, often arranged in extra rows; a full and fluffy flower.

drip line – 1) The leaves of trees (and some smaller plants) with fibrous roots generally have leaves that droop down carrying water to the outer margin of the roots. This moistened circle is known as the drip line, and indicates a change in microclimate. The area under the tree has less rain, sun and wind, so plants have to compete with the tree's roots. See edge.

drought (dr out) – Severe lack of moisture, water stress.

drupe (droop)- A fleshy fruit with a single stony seed (endocarp), like a cherry, *Prunus*.

drupelet (droop'let) - A tiny fleshy fruit with a hard center (endocarp), often joined in a cluster (aggregate fruit), like a raspberry, *Rubus*.

DYD (dee-why-dee)- Durned Yellow Daisies. The yellow members of the daisy family, Asteraceae (Compositae), are a huge group and the florets of most of them are tiny. They also tend to hybridize, so are difficult to tell apart without a lot of lens or microscope work on both the flower and fruit. If you ask for help on the identification, you are likely to be told you have a DYD, so it gets listed as another unknown.

E

e- (ee, eh) - A prefix meaning without something, as eciliate means without eyelash-like hairs (cilia) and eglandular means without glands.

ebeneous (ee ben' ee us) – Black

eccentric (eks sent' trick) – Just a hair off center. See ObFred.

echinate (eck in ate) – Prickly as a hedgehog.

economic botany – Figuring out how to make money from plants. George Washington Carver was a genius at this, finding over 300 marketable products from peanuts.

ecostate (ee cost' ate) – Without a midrib.

ecosystem (eek' oh sis' tem) – The community of plants, animals and micro-organisms that inhabit the chemical and physical environment of an area.

ecotone (eek' oh tone) – Transition zone where one plant community abuts another, like the timber line on a mountain.

ecotype (eek' oh type) – A group of plants that have adapted to a certain habitat or set of conditions.

ecto- (eck' toe) A prefix meaning outside of, external. For example, in a cherry, *Prunus*, the **ectocarp** is the skin, the **mesocarp** is the meat, and the **endocarp** is the stone containing the embryo. The entire fruit wall (ectocarp and mesocarp and endocarp) is a **pericarp**.

edentate (ee dent' ate) - Without teeth.

edge – The meeting line of two distinct habitats, like a marsh and a meadow, or a woodland and a river. Edges produce greater variety of species of plants and animals.

edged - Having a very narrow band of color around another color. See bordered, picot.

edible (ed' ih bul) - May safely be eaten. Many fruits are said to be edible. The safest way to tell: look for a cluster inside a plastic shell with a transparent top and a rubber band around it in the habitat of the produce section of the grocery store.

effuse (eff fyoos') – Spilling freely; looser than diffuse.

eglandular (ee gland' you lar) - Without glands.

elaminate (eh lam'in ate) – Referring to a modified leaf without the green part (blade), such as a cactus spine or grape tendril, *Vitis*.

elliptic (ell lip' tick) – Shaped like a tangerine or the Earth, broad in the middle, narrower near the poles; a flattened oval.

elongate (ell long' gate) - Stretched out, longer than is common.

emarginate (em marj' in ate)- With a little notch instead of a point where the tip of a leaf or petal should be.

embryo (em' bree oh) - The baby (rudimentary) plant inside the seed; the germ. Synonym: zygote.

emergent (em merge' gent) - With part of the plant underwater (submersed), and the emergent part rising above the water into the air.

emersed (em merst') - Raised above the water. Antonym: submersed.

end bud - A bud at the end of a twig. See apical bud, terminal bud.

endemic (end em' ick)- 1) In botany, this means that a plant grows only in a certain geographical area where conditions are perfect for it, will not transplant or grow well from seed elsewhere.
2) In general usage, it is loosely used to mean native.

endo- (end' oh) - A combining form meaning internal, the one on the inside. See ecto- for an example.

endocarp (end' oh karp) - The innermost layer of the fruit wall (pericarp) containing the baby plant (embryo), like a cherry pit, *Prunus*. See ecto- for example.

Engler-Prantl – The natural system of classification developed by Adolph Engler (1844-1930) and Karl AE Prantl (1849-1893) that was used in most botany books during the 20th Century, at least until the 1970s.

ensiform (en' sih form) - Sword-shaped, as the leaves of wild flags, *Iris*.

entire - With a smooth margin, without teeth or divisions.

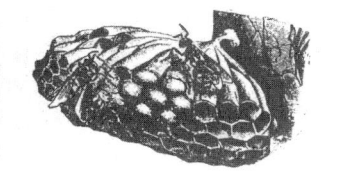

entomophilous (en' toe moff' fill us) – Pollinated by insects.

envelope - The surrounding part of anything. In botany, usually refers to the floral envelope, that is, the petal and sepal coverings of the essential organs (stamens and pistil).

epetiolate (eh pet' ee ole late) - Said of a leaf without a stalk (petiole). Synonym: sessile leaf.

ephemeral (eh fem' er al) - 1) Germinating, growing, flowering and fruiting in a short period of time, as desert plants following a rain; or woodland spring wildflowers.
2) Lasting a day or less.

epi- (ep' ih) – 1) A prefix which means on top of, as an epiphyte is a plant which grows on top of another plant without being a parasite, just using the host for support.
2) May mean beside, among, on the outside, anterior; as an epicalyx is a circle of growths (involucre) that looks like a second calyx.

epipetalous (ep' ih pet' al us) - Said of stamens (anthers) which are fastened directly to the petal without a filament.

epiphyte, epiphytic (ep' ih fight, ep' ih fit' ick) - Growing on another plant, but not as a parasite, as when a moss grows on the side of a tree (yes, moss grows on the north side, and the east, west and south sides...).

equinoctial (ee kwin knock' shall) - Having flowers that open and close at certain hours of the day, as in four-o-clocks, *Miribilis*.

erect - Standing up straight, not flopping over.

erose (air roh'ss) - Leaf margin appearing gnawed by a mouse, with shallow, irregular teeth and lobes. **Lacerate** is similar but with much coarser sections. **Incised** is like lacerate, but cut more deeply and narrowly. **Laciniate** is cut into ribbons with the deepest and narrowest 'leaflets'. See appendix: leaf - margins.

erosion (ear rose' shun) – The process of soil being carried away by wind or water. See French drain.

escape - An introduced (alien, exotic) plant that has become naturalized. If a student leaps out of bed and runs wild, he is an escapee, but when a plant strikes out from its bed and runs wild, the correct use of what is normally a verb becomes a noun.

-escent (ess' sent) – Combining form meaning inclined to be.

essential flower organs - The male parts (stamens) and female parts (pistils) necessary for fertilization and fruiting. Synonym: reproductive organs.

evanescent (ev' an ess' cent) - Disappearing early or quickly. See ephemeral.

even-pinnate - A compound leaf with the leaflets opposite each other on the leaf stalk (petiole, rachis) and no leaflet at the end of the stalk. Synonym: paripinnately compound. See appendix: Leaf - arrangement.

evergreen - 1) In the North, this usually refers to pine and pine-like (coniferous) trees.
2) In the South, any plant that stays green all year without dropping its leaves.

everlasting – Flowers with papery petals that can easily be dried for winter decorations or art projects. Over 30 genera/species carry the name, and others like goldenrod, *Solidago*, hold their color well.

excrescence (eks cress' sense) - An outgrowth from the surface.

exocarp (ecks'oh carp) - The outside layer of the fruit wall (pericarp). Synonyms: ectocarp, epicarp. See ecto-.

exotic (eck zot'ik) - From another country, brought in on purpose or by accident. Synonym: alien.

exserted (ecks zert' ed) - Sticking out of, as male organs (stamens) much longer than the petals, so they extend beyond the flower. Synonyms: protruding, stretching.

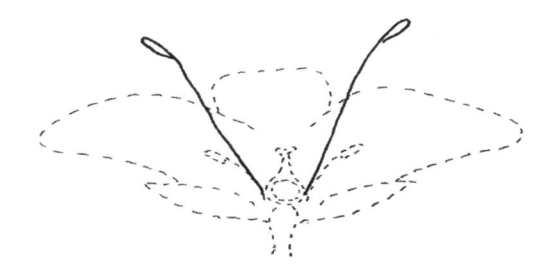

exudate (eck'zoo date) - Any liquid discharged through openings, as though sweated out. Synonym: ooze.

eye – 1) The different colored center of a flower.
2) An undeveloped growth bud, as on a potato.
3) What the boys in class give a new girl.

F

fairy ring – A circle of mushrooms or toadstools. Legend claims fairies dance within the circle. See fungus, mycelium.

falcate, falciform (fall' kate, fall' sih form)- Strongly curved and tapering gradually, like the wings of a falcon or blade of a boomerang.

family – One layer of taxonomic groups (classification) between order and genus. See taxon.

farinaceous (fair' in ace' ee us)- With a surface that feels like somebody spilled fine corn meal on it.

fascicle (fass'sickle)- A cluster of something, as roots of spiderwort, *Tradescantia*, or needles of white pine, *Pinus strobus*.

fastigiate (fast tij' ee ate) – With branches that swing upward until almost parallel with the tree trunk or stem.

fauna – 1) All animals in an area from bees to buffalo.
2) An exchange student studying botany.

felted - With thick matted hairs.

female - 1) Producing eggs (ovules).
2) Referring to a plant with only female organs (pistils) in the flower or on the plant. Synonym: pistillate.
3) A politically correct term that will make girls happier to help you identify plants.

fence - A wire barrier to keep out unwanted animals like botany students.

fenestrate (fen' eh straight) – With tiny windows, either holes or translucent membranes.

fern – 1) A cryptogam, not a flowering plant.
2) The countries that faunas came from.

fertile - With male (stamen) or female (pistil) parts, or both. Ready to reproduce.

fibrous - Having loose woody cords, as the outer layers in some fruits such as wild cucumber, *Echinocystis*.

fibrous roots (fie' brus)- Common roots, made of masses of slender shoots, sort of like a bushy plant upside down. Compare tap root.

filament (fill' ah ment)- 1) The stalk of a stamen, the sterile support of anthers.
2) Any thread-like body
3) A question to another student following a lecture as, "Do you know what the filament?"

filantherous (fill an' ther us) - Referring to the most common form of male organ (stamen) in plants, made up of a filament and two anthers.

filiform (fill' ih form)- Like a thread; long, slender and usually round in cross section.

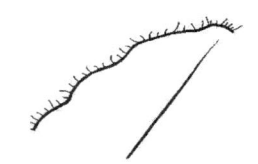

fimbriate (fim' bree ate)- 1) Fringed.
2) With little eyelashes, as a stigma. Synonym: ciliate.

fistulose (fist' you lohs) – With a hollow passage, as a tube or pipe.

flabellate (flab' bell ate) – Spread open like a fan.

flaccid (flass' id)- Limp and droopy, not rigid.

flange (flanj) – A ledge or projecting rim.

flat – 1) A shallow tray to start cuttings or seedlings.
2) What your car gets driving down side roads looking for new species.

fleshy roots - Succulent, juicy roots, not woody, sometimes edible.

flexuous, flexuose (fleck' shoe us, fleck' shoe ose) - With a series of curves along a line (axis); the way you hope any snakes may look heading off in the other direction. Synonym: zigzag.

floating - Growing on the surface of water, like duckweed, *Lemma*.

floc (flock) – Soft wooly hairs in a cluster.

forcing – 1) The process of producing a flower before its usual season.
2) The process of provoking memory to recall botanical terms.

flora – 1) A collective term to cover all the plants in an area.
2) A book focused on plants of a certain geographical area, like a state, a mountain range, or a particular lakeshore.

floral envelope – Sepals and petals surrounding the essential organs (stamens and pistils). Synonym: Accessory organs. See perianth.

floret (floor' et)- One tiny united flower clustered in a whole mess of florets that make up a blossom head, seen in a daisy, aster or sunflower, Asteraceae (Compositae). See ray floret, disc floret.

fluted (flew' ted) – With grooves and ridges.

foliaceous (foal' ee ay' sus) - Said of a flat part of something other than a leaf, with the color and texture of a leaf, like the immature fruit-bracts of hops, *Humulus*.

follicle (foll' ih kull)- A dry fruit from a single female part (carpel) with many seeds, that opens with only one slit (suture) along the side, as in milkweed, *Asclepias*.

forb, forbe (four b) - 1) Any non-woody green (herbaceous) plant which is not grass-like (monocot).
2) In some areas of the Southwest US, the term refers only to forage plants, i.e., non-grassy plants eaten by livestock.

forestry – The scientific practice of planting and managing trees to produce the best and most wood.

frass - A word that polite people use for the by-product of bugs that got into your herbarium specimens.

free - Separate, not fastened to similar organs, not united. Back of flower shown with free petals and free sepals.
Synonym: Distinct.

free-petal - Referring to a flower where none of the petals are fastened to each other, just fastened to the base (receptacle, hypanthium, ovary, pedicel).

French drain – An inexpensive, attractive, convenient erosion control method: Dig a trench crosswise of the slope being eroded; fill the trench with stones; smaller stones on top. Water moving downhill is trapped and slowly released into the water table.

frequency - 1) Rarity or abundance of a species.
2) A general term referring to the relative number of times group members developed poison ivy rash.

friable (fry' ah bull) – Describes soil that crumbles easily in your hand.

fringe – 1) A non-technical term for little hairs on the edge of a leaf or petal. See fimbriate, ciliate.
2) Where you get pushed until you learn to talk Botanese.

frost pocket – A geological basin surrounded by hills where cold air flows down and cannot escape, so it remains colder than the higher area. See microclimate.

frosted - Referring to a surface that has a waxy coating. Synonyms: pruinose, sebiferous.

fructiferous (fruck tiff' er us) – Capable of producing fruit.

fruit – 1) The matured ovary of a plant, normally with seeds, with or without extra (accessory) parts.
2) Cooks belong to a different subculture, and have a different vocabulary. In food preparation, "fruit" refers to any sweet or sweetened vegetable part, usually part of a ripened pistil served as dessert, but includes the stems of rhubarb, *Rheum*, and the large stem-like syncarps of pineapple, *Ananas*. "Vegetables" to cooks are lightly salted, raw or cooked, usually stems, roots or leaves, but many are botanical fruits, like beans, *Phaseolus*; tomatoes, *Solanum*; and olives, *Olea*.
3) The seed-containing part of a plant, necessary for correct botanical identification, that splits, rots, falls off, hasn't developed yet or has been eaten by birds when you attempt to make the aforesaid identification.

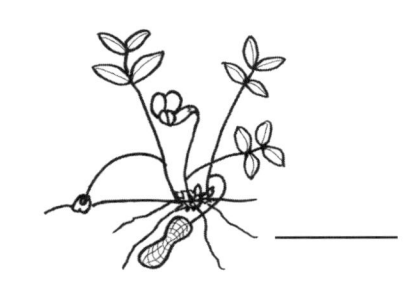

It may be hiding anywhere on a plant including among the roots, as found on the peanut, *Arachnis hypogaea*, on which the blossom is above ground, but penetrates the soil after pollination and ripens there.

frutescent (fruit ess' sent) – Nothing to do with fruit, simply means woody with many stems, bushy.

full shade – Not a reference to a ghost who had a large lunch, just means that the plant does not do well in sunshine. The interior of a deep forest provides full shade. So few full shade plants exist that the deep forest floor might be called a biological desert.

full sun – Nothing like the full moon, just means the plant should be in sunshine for a minimum of 6 hours a day.

fungus, plural **fungi** (fun' gus, fun' guy) – Plants with no flowers, some edible, some fatal. People who study them are called mycologists. See mycologist, cryptogam.

funnel - A tube flower that is wider at the open end, with all petals joined together for nearly full length, such as a morning glory, *Convolulus*.

furcate (fur' kate) – Forked.

furrow - 1) A groove.
2) The line that forms on the forehead of people identifying plants.

fusiform (fuse' ih form) – A long slim part, thick in the middle with narrow ends.

G

gall (gawl) - 1) A swollen area on a plant caused by an insect or disease.
2) What a second year botany student may have a lot of that can result in a swollen area on his nose.

genotype (gee'no type) – The traits carried by inheritance factors (genes) of a plant (or animal) that may or may not show in the appearance (phenotype) of the thing. See phenotype, Mendelian law.

genus, pl. **genera** (jee' nus, jen' er ah)- The first and capitalized part of a scientific name (binomial). A grouping of similar plants, classified within a family, and including several related plants known as species. See 'binomial' for more complete explanation. See taxon.

geotropism (gee ott' row pism) – A plant's response to gravity. Roots that grow down have positive geotropism, while limbs that grow up exhibit negative geotropism.

germination (germ' in ay' shun) – When a seed breaks dormancy with the right combination of moisture, warmth and light to produce active growth.

gibbous (gib' us)- Ballooned on one side, like the bottom of the flower of ——————— jewelweed, *Impatiens capensis*.

gin (jin) - 1) A helpful farm invention for separating parts of plants.
2) A helpful distillery invention for botany students separating species.

girdling – 1) In reference to animals chewing off the bark of a tree or shrub for starvation food during the winter.
2) The strangling of a woody stalk by inflexible materials like wire or rope put there for planting support or to hold a swing on a mature tree.
3) What women forget while gardening.

glabrate or **glabrescent** (glay' brate, glay bress' sent - most people verbally skim over 'glay' so it comes out almost glab' rate)- With a hairless (glabrous) surface, or the surface becoming hairless as in some plants where the young stems are hairy and the hairs drop off at maturity.

glabrous (glay' brus)- Smooth, without any hairs on the surface. Compare with blistered, like your skin after you attempt to revive the campfire with gasoline.

glade (glayd) – A woodland opening. This may be temporary, as a hole left in the canopy when an old tree fell, or more permanent because of a rocky outcropping or other factor that interferes with growth. See canopy.

gland - A tiny lump, dot, or pit found on the surface of an organ. It may or may not secrete something such as farina, nectar or resin.

glandular (gland' you lar)- Referring to a surface with many tiny glands, varying from blackish to clear, usually producing sticky or oily liquid.

glaucous (glaw' cuss)- 1) Having 'bloom', covered with a powder or a smooth, waxy coating that rubs off, like a blueberry, *Vaccinium*.
2) bluish or whitish, with a frosty appearance.

globose (glow' bose, glob' ose) - 1) Shaped like a sphere.
2) In flowers, united petals forming a sphere with a hole at one end for bees to stick their noses in. Compare with subglobose.

glochid (glah'kid) – A nearly invisible barbed prickle found on Cacti, that you are unaware of until it penetrates your finger. It proves that those who say they love the genus are truly perverted S&M freaks.

glomerate (glom'er ate) – Crowded, densely clustered.

glossary – A book with terrific vocabulary but a weak plot.

glutinous (glue' tin us) - With a sticky coating.

GPS (gee pea ess) – Global Positioning System. A hand-held tool that allows you to accurately relocate where you have been, perhaps to later collect seeds of an admired or unknown plant.

grade - 1) The degree and direction of slope of a hill.
2) The competition goal that can take the excitement out of learning.

grafting – 1) A method of taking a part (scion) from a desirable, often slow growing plant, and attaching it to a cheaper, older, deep rooted tree (stock) to grow better and faster. I asked my grandfather if there was any time you couldn't do a graft: "Yes, when your jackknife is dull."
2) While working in a highly paid government job as a consultant, I purchased Garner's excellent *Grafter's Handbook*. Seeing it on my desk, my boss told me to put it in the drawer and keep it there.

grain - A dry fruit with one cavity (locule) where the seed coat sticks to the wall of a fruit that does not open at maturity (indehiscent). Many are found in the grasses, Poaceae (Graminae).

grandi- (grand' ee; gron' dee) – Combining form meaning large.

granular, granulous (gran' you ler, gran' you luss)- A surface that is somewhat mealy, but less so than that called farinaceous.

grass – 1) A plant of the Poaceae (Graminae) family. People often think one grass is best for the lawn. **Warm season grass** does not green up during short winter days even in the South, and **cool weather grasses** turn brown or even burn out in summer heat. A mix is best.

2) Forage plants, green herbage furnishing food for livestock, either pasture or hay, that may include forbes.

3) Loosely, and incorrectly, any plant with long slender leaves, such as reeds, sedges, etc.

4) Dried plants that some students study intensely on weekends.

green cuttings – Young branches that contain growth material (meristem). See new wood.

gregarious (greg' air ee us) – 1) Found in clumps or colonies.

2) How you feel after a long field trip and someone mentions your favorite bar.

ground cover – Plants other than grass that provide a low-growing carpet between taller plants.

ground nut - 1) Apios americana, the potato bean.

2) A person who insists on classifying soils as part of the plant identification process, and who has been known to study for a soil test.

growing point – Tip of a root or stem where it lengthens. See meristem.

growing season – The part of the year from the first leaf showing on a plant until frost or lack of light either kills the plant or forces it to go dormant.

grub - 1) Beetle larva.

2) Your plant specimens from a beetle's point of view.

gymnosperm (jim' no sperm) – A plant with seeds naked rather than inside an ovary, like the pines (conifers).

gynoecium (jine knee' see um)- All the female parts of a flower; the pistil, especially to explain several female parts (carpels) as a single unit.

H

habit - The way a species likes to grow. General appearance as to whether the complete plant is upright or floppy, twiny or stiff, etc.

habitat – The type of ecological niche that a plant prefers, such as swamp, desert, forest, plain, etc.

half hardy – 1) In reference to outdoor perennials, a plant that cannot withstand the cold and wet of winter and will die in repeated frosts.

2) An indoor plant that can stand temperatures of 50 to 55 degrees F. for healthy growth. See tender and hardy.

halophyte (hal' oh fight) – A plant that can survive with a large amount of salt (NaCl) in the soil.

hamate (hey'mate) – Hooked, especially at the tip (apex).

haplo- (hap' low) – Combining form meaning single or simple.

haploid (hap' lloyd) – With a single set of chromosomes. Natural, not hybrid.

hard science – Those disciplines like math, biology, and chemistry, where you need to really hit the books with lots of memory crutches, since your answers can be verified as right or wrong after aspects are investigated and hypotheses proven or disproved. See soft science.

hardiness, hardy – The ability of a plant to withstand cold or other factors. 1) For outdoor perennials, those sold as hardy plants easily live through the cold and wet of winter.
2) For indoor plants, those that withstand temperatures of 45 F or lower are termed hardy.
3) The term is occasionally used for plants that survive in hot dry (arid) areas as summer hardy. See zone.

hastate (hass' state)- Shaped like an arrowhead, but with the bottom corners pointing out or up.

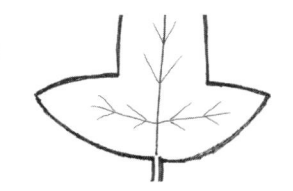

head - A compact cluster of flowers or florets with no flower stalks (sessile florets). The cluster appears to be one flower, like a dandelion, *Taraxacum*, or clover, *Trifolium*.

hedge – A row of dense or thorny plants nearly impossible to penetrate. People usually have planted them for privacy or windbreak, but they may appear naturally on the edge where a woodland meets a meadow. A **tapestry hedge** contains several species of flowering shrubs.

heel – 1) A grafting term for a strip of bark and wood with a bud or shoot extending, cut from a branch to transfer to another woody plant.
2) A greenhouse term for a side shoot cutting with stem tissue attached to improve chances of rooting.

heeling in – A process to hold transplants for variable times by digging a trench, laying the transplants in at an angle crowded a few inches apart, generously covering the roots with moist soil to keep them from drying out, then pressing firmly with the heel of your boot. A week or month, (or - for impulse buyers and gift recipients who haven't a clue where to plant them - a year) later, the plants are carefully dug and planted where they are to remain.

helicoid (heel' ih koid) – Coiled like a snail shell.

heliotropic (he' lee oh trop' ik) – Response of plant parts to a light source, like a flower face following the sun as it moves across the sky.

herb (Americans say erb, English say herb)- 1) Any green plant that dies back each year, without woody stems above ground.
2) A plant used for flavoring or medicine.

herbaceous (her bay' shus)- 1) Referring to a plant that is not woody above ground and dies back in winter.
2) Describing any relatively small green plant, including grasses.

herbarium, plural **herbaria** (her bare' ee um, her bar' ee ah) –
1) A place with carefully controlled moisture, light and heat where labeled, mounted, dried plant specimens are kept, sometimes including wood and seed collections. A reference library of pressed plants.
2) A place where botany students are held prisoner until their sentence (or paragraph or essay or thesis) is completed.

herbarium sheet - A heavyweight paper of a standard size on which dried, pressed plant samples are mounted for study and comparison.

herbivore (her' biv ore) – An animal that feeds exclusively on plants, i.e., deer, cattle, sheep.

hibernal (high burn' ul) – Flowering or growing in winter.

high shade – An area where lower branches of trees were removed, perhaps to provide short sunshine for base plantings, to open a 'window ' to a view, for a horseback riding trail, or to allow vehicles to gain access to back country. Other names for this are **limbing up**, or **raising the crown.** Road crews have been known to do this on roadside trees for a homeowner if a case of beer appears.

hilum (high' lum) – A scar on a seed where it had been fastened. See appendix.

hip - 1) An accessory fruit with the widened stem tip (receptacle) and/or floral base (hypanthium) formed into an urn shape, with the small hard seeds (achenes) inside, as in a rose hip, *Rosa*.
2) The widest part of a botanist's anatomy from all the hours spent captive in the herbarium.

hirsute (her suit')- Covered with long stiff hairs - softer than **hispid**, but firmer than **hoary** or **pubescent -** as seen on squash vines, *Cucurbita*.

hispid (hiss' pid)- With very long, stiff, sharp hairs. Bristly. Seen on the fruits of teasel, *Dipsacus*.

hoary (hoe' ree) - 1) With soft short white hairs, found in the leaves of catnip, *Nepeta* and horehound, *Marrubium*.
2) Grayish white.

hoe – 1) A gardening tool sometimes taken on a field trip for collecting.
2) Dropped on the ground and stepped on by a fellow hiker, he identifies the tool by name when the handle hits him in the head.

hollow pith – Referring to stems that are empty between the stem joints (nodes). See appendix – stem - pith.

honeydew – Globs of sweet sticky stuff left on plants by insects like aphids and whiteflies.

hood - 1) A cup-shaped petal over a flower, as monk's hood, *Aconitum*.
2) A covering over the mouth of a capsule.

Hort – Written following a scientific name where the author's name should appear, Hort (with or without a period) means that the garden plant is known and sold by this name in common usage, but has not been described and formally approved by this name.

horticulture (hore' tick culture) – 1) The scientific study of gardening; growing fruits, vegetables, flowers and ornamental plants, with the goal of improvement. Compare with agriculture.
2) The complicated study of people growing plants, e.g., you can lead a horticulture but you can't make her learn.

horticultural therapy – Professional big words that mean taking those who got lost in the concrete jungle until scared silly, into a place where they can reach out and combine healthy soil and fertile seeds; letting them get their hands dirty and their minds clean.

host – 1) A living plant providing food and support for a parasitic plant.
2) Sometimes refers to a strong plant that provides support for a vine.
3) The neophyte botany student who is trying to get more experienced students to help him. See bar.

humifuse (hyoom' ih fuse) – Spreading over the ground, like a cucumber vine, *Cucurbit*.

humus (hew' muss) – Decayed vegetation; the process of decay in vegetation. An important factor of healthy soil.

hyaline (high' ah lean) – Usually said of membranes that are thin and translucent or transparent.

hydrophilous (high drof' fill us) – Pollination with pollen being carried by water to the female organ (pistil).

hydrophytic (high' dro fit' tic) – Adapted to growing in wet places.

hypanthium (high pan' thee um) - A flower part shaped like a ring, cup or tube. It appears that the sepals, petals and stamens are fastened to this, but actually the bases of those organs are fused together to form it. Found in many roses, *Rosa*. Synonyms: floral tube, floral cup, calyx tube.

I

illegitimate (ill' ledge it' ih mat) – 1) Not an accepted name for a species, breaking Articles of the International Code of Botanical Nomenclature.
2) In reference to your paying someone to collect and identify your required number of species for botany class.

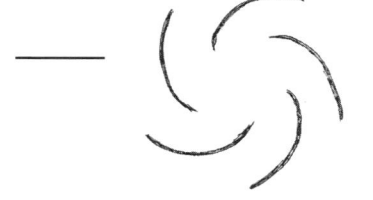

imbricate (im' brick kate)- Layered and overlapping like shingles. While this pattern shows in many flowers, it is most common in buds, with petals overlapping as shown in this cross-section.

imperfect - With no male parts (stamens), or alternately, no female parts (pistil), in a flower, leaving it with one sex only, but still fertile. Synonym: unisexual. Antonyms: perfect, bisexual.

implicate (im' plick kate) – Twisted together, entwined.

incised (in sighs'd)- Cut into uneven sections, especially sharply and repeatedly but irregularly, as a leaf margin, more coarsely than 'erose', but not evenly enough spaced to be called 'lobed'.

inclined - Angling upward. See ascending for comparison.

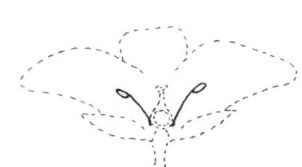

included - Not sticking out, as might describe the stamens staying below the tops of the petals. See exserted, inserted.

incurved – Describing anything that curls toward the center instead of outward, as the tip of a leaf curving upward toward the attachment, somewhat covering the top of the leaf; like your tongue when you try to touch your nose with it. Opposite of recurved.

indehiscent (in dee hiss' sent)- 1) A fruit that doesn't open by itself, such as a nut.
2) Not to be confused with 'indecent', a nut that does open itself in public.

indeterminate (in dee term' in at)- Referring to a cluster of flowers (inflorescence) that starts blooming on the bottom of a stalk or the outside of a cluster, while the central stalk (peduncle/rachis) continues to grow as the blossoming moves upward or inward. See determinate.

indigenous (in dij' en us)- Native to that part of the country, not alien. Not to be confused with indigent, which is depauperate, poor.

indurate (in ' doer at) – hard, stiffened.

indusium (induce' ee um) – An outgrowth under a fern leaf that protects the spores.

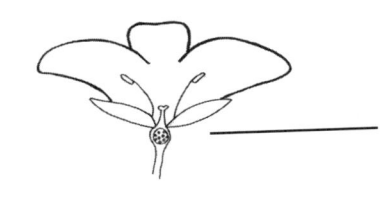

inferior - Merely means below, not referring to quality or function, as your feet are inferior to your legs. An inferior ovary is completely functional, is simply located below the petals. Antonym: superior. To be superior to that botanist who makes you feel dumb, just go to the top of the building where his office is located, and you will be spatially superior to him.

inflated – 1) Describing parts blown up like a balloon, bladder-like.
2) A good description of many botanical egos.

inflexed (in ' flecks d) – Bent inwards. See reflexed.

inflorescence (in floor ess' sense)- 1) The pattern in which flowers grow on the plant.
2) The coming of a plant into flower.
3) The flowering part of the plant.

infundibular (in fun dib' you lar)- Funnel shaped, like the corolla of some united petal flowers.

insect repellant - A solution required to be sprayed all over on field trips. It rarely keeps bugs away but other students give you room to work when you return to the lab.

insectivorous plants (in' sect iv' or us) – Plants that live in conditions where nitrogen is unavailable from the soil so they capture and digest insects whose flesh provides the nutrient. Synonym: carnivorous plants.

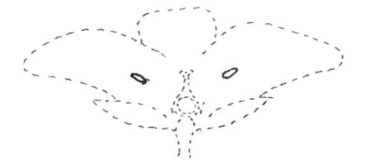

inserted - Stuck into. Said of stamens (anthers) that are fastened directly to the petals with no filament; or similarly, any small part growing from an unusual place.

integrated pest management – A buzz phrase used by the in-crowd to mean biological pest control. Called IPM.

interfoliar (in ter foal' ee er) - Between the leaves. Said of flowers that are not at the ends of stems but tucked down into the foliage, such as those found in most mints, *Mentha*. Synonym: axillary.

International Code of Botanical Nomenclature (no' men klayt' your) - The legal articles that control proper description and order of plant names (taxa).

internode (in' ter node)- The stem section between the joints (nodes) of a stem.

interrupted – Not continuous, has gaps in the line.

invader – 1) A weedy plant that grows where most plants cannot survive. In normal habitats, it often crowds out more desirable plants. Often an alien. Synonyms: weed, pioneer plant.
2) The inflated ego who hops into your car uninvited on a field trip.

involucel (in vol' you sell)- 1) A secondary group of leaflets as on the individual flower clusters (umblets) of a compound flower cluster (umbel) like Queen Anne's Lace, *Daucus carota*.
2) A second group of bracts below the sepals (calyx), as in Wine Cups, *Callirhoe*. See appendix.

involucre (in vole luke' er)- A grouping of many bracts/ leaflets (phyllaries) around the bottom of an aster type flower, Asteraceae (Compositae), covering the enlarged stem tip (receptacle). See appendix – flower – looks like sepals but aren't.

involute (in' vol lute)– Having side margins rolled upward over the top surface, the way some people can roll their tongue from the sides. Compare with revolute.

IPM (eye pea em) – Integrated Pest Management. A labor intensive program that encouragers farmers and other growers to monitor fields, especially monoculture, for disease and insects, treating only the infested areas, not the entire field. Chemicals are kept to a minimum and natural insect predators are extensively used. See monoculture, organic gardening/farming, crop rotation, BMP.

irregular flower - A flower that is not mirror image in all directions, with petals not evenly distributed. Synonyms: zygomorphic, asymmetrical. Antonyms: actinomorphic, regular, symmetrical.

J

Jack-in-the-pulpit - 1) *Arisaema triphyllum*.
2) A politically incorrect name denoting religion, which is in the process of being renamed Ralph-in-the-reformatory.

jointed - With swollen nodes on the stem. The stems often form a zigzag pattern, bending at the swollen nodes.

jugate (jew' gate) – In pairs, overlapping.

junciform (jun' see form) – Looking like a rush.

K

keel – 1) The line where the two wing petals meet in legume flowers.
2) A petal or leaf with a V-shaped cross section like the bottom of a boat.

kernel (kur' nel) – The edible part of a nut, a nutmeat.

key - 1) A long slender fruit that has a globe on one end, with a wing growing off it, resulting in a new tree when dropped to the forest floor. Synonym: samara.
2) A shiny object consisting of a round terminus and a winglike structure having one serrated margin, dropped to the forest floor during a field trip, resulting in a broken car window to reach the cell phone to call a locksmith.
3) **floral key** - A device to help you identify a plant by giving you a series of choices. Each choice that you make leads to another choice. Soon you reach a choice regarding a structure or quality not present on the specimen in front of you, so you leaf through all the pictures of a brightly illustrated non-technical book to make a 'positive' identification.

L

labiate (lob' ee ate) – In reference to a flower with lips.

lacerate (lass' er ate)- 1) Appearing torn, as a leaf margin, more deeply notched than in erose.
2) laceration - n. What—on a field trip—you tend to collect more of than specimens.

laciniate (lah sin' ee at)- Referring to leaf margins appearing slashed into ribbons with narrow points almost like a fern. See erose.

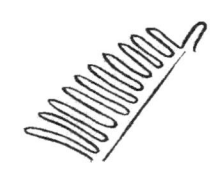

lacustrian (lack coo' stree an) – Grows around lakes.

lamina (lam' in ah) – Any wide flat surface of a plant that captures light for photosynthesis; leaves.

lanate (lan' ate) – With woolly hairs.

lanceolate (lance' ee oh late)- Having an outline shaped like a lance point or arrowhead, with the wider part closest to where it is attached.

When the wide part is farthest from the attachment, it is called **oblanceolate.**

lanulose (lan' you lows) – With short woolly hairs.

larva (lar' vah) – The worm stage of a butterfly, moth or beetle.

latent (lay' tent) – Concealed, not clearly seen.

lateral (lat' er ull)- At the side, as a lateral branch growing out from a stem.

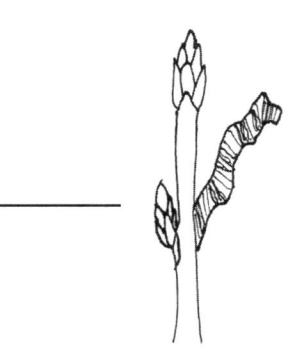

lateral bud - A bud that forms on the side of a branch at the base of the leaf stalk (petiole). Synonym: axillary bud.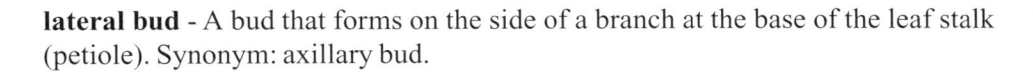

lax – Not upright, floppy.

leaching (lee' ching) – A process set in motion by over-watering, either natural or from your garden hose. The movement of water through the soil washes away substances, more so with porous soils like sand. Leaching may be helpful if the soil contains high amounts of something harmful like salt(s), or be harmful itself if necessary nutrients are lost.

leader – 1) The main stem (axis) of a woody seedling or sapling, that will eventually become the trunk of the tree.
2) The part of a tree reaching for the sky.

leaf - The organ of a plant that breathes and makes food from sunshine, minerals and water. The most recyclable part of a plant, falling to the ground to produce nutrition to feed the plant to grow new leaves to fall to the ground, etc. It consists of two parts, the blade and leaf stalk (petiole).

leaflet – One of the blades of a compound leaf.

leaf scar - A mark on the bark of a tree or shrub that shows where a fallen leaf had been attached. This makes winter identification of trees quite easy with a good winter field guide. Inside the leaf scar, you will see little dots that are the **vein scars,** also called **leaf trace.** The number and arrangement of these, combined with the shape of the leaf scar and outer bud scales allows winter identification.

legend - 1) The text under a picture or map.
2) Story from long ago.
3) What you become after five various mishaps in a single field trip.

leggy - A non-technical term referring to a plant that has lost most of its lower leaves but retains a tuft of leaves at the top, so it looks more like a palm tree that what it really is.

legume (leg' yoom, leg goom')- 1) A member of the Leguminosae family, such as pea, *Physalis*, and bean, *Phaseolus*. These plants display an unusual flower (papilionaceous).

2) The pod, opening on two sides with seeds fastened to pod, found in that family.

3) Edible seeds of the family.

lenticel (lent' ih sell) - A small, corky dot or streak on the bark of trees and shrubs, arising normally around pores in the bark; sometimes on roots or fruits.

lentiginous (lent ij' in us) - With crusts of scales that rub off. Synonym: scurfy.

liana (lee an' ah) – 1) Any climbing plant that roots in the ground. The term is used mostly for rain forest vines, which are nearly all woody (ligneous).
2) A term known to all crossword puzzle fans.

lichen (lye' ken) – A small scratchy growth composed of both algae and fungus that grows in patches on inhospitable stones or soil breaking them down for use by other plants. Some look almost like regular plants.

light dependent seed – A seed that has to be exposed to light to germinate, usually from a plant whose seeds germinate immediately on touching soil, as compared to those who wait for the following spring. See dark dependent seed.

ligneous (lig' knee us) – Containing woody cells, as opposed to soft herbaceous plants. In reference to trees, shrubs and woody vines.

ligulate (lig' you late)- 1) Strap-shaped, like the "petals" of the outer ring (ray florets) on a daisy, aster, or sunflower. Synonym: lingulate.
2) Having a tongue-like projection. See strap flower.
3) In reference to a floral head that contains only ray florets, no disk florets.

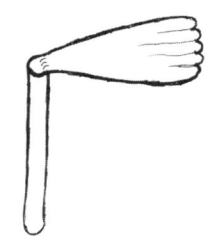

limb - 1) The open part (blade) of a leaf, or the strap of a floret.
2) Non-technical term for a branch of a tree or other plant.

limbing up – Removing lower branches of a tree. See high shade.

linear (lin'ee ar)- Forming ridges, long and narrow, as found in leaves of grass, Graminae.

Linnaeus – Carolus Linnaeus, born Carl von Linne, proved a botanical genius. The Latin spelling was granted in honor of his knowledge. He invented the binomial nomenclature system with two word names for plants in the 1700s when a teenager. It remains in use today.

lingulate (ling' gyou late) - Shaped like a tongue. Synonym: ligulate.

lips - The petals on a flower that look like a mouth (bilabiate), as found in many mints, Labiatae (Laminae).

littoral (lit' or al) – Grows along shores. See riparian.

loam (low m) – A high quality soil made of roughly equal parts of sand, silt, clay and well rotted plant matter.

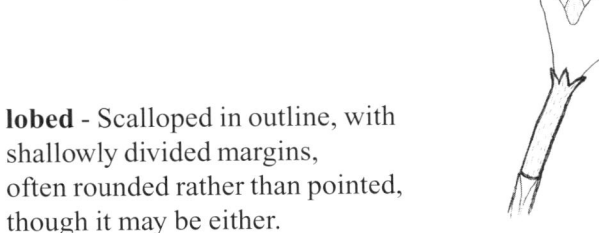

lobed - Scalloped in outline, with shallowly divided margins, often rounded rather than pointed, though it may be either.

locule (lock' yule) – The inside cavity of an ovary that holds the ovules.

lodging – In reference to plants, especially cereal grains like oats and wheat that are close to harvest, being flattened by wind and rain.

loment (low' meant) – A legume pod that nips in between each seed.

longitudinal (lonj' ih tood' in al) – Situated the long way, like grooves.

loose, distant, scattered - Interchangeable terms meaning that parts are arranged far apart, usually with little or no pattern. Synonym: sprawling.

lumber – Wood that has been cut to size and prepared for use in building. Misuse of the word is so common dictionaries now list timber and lumber as synonyms; don't make that mistake with foresters, woodcutters and lumberjacks. See timber.

lumpers and splitters - Lumpers are botanists who believe that each species of plant has a lot of variations when it grows in different habitats and climates, but they remain the same species. Splitters think that every variation should have a different species name, and then break those down into subspecies. Don't worry. Be happy, even if you only identified the genus. You might better correctly list *Solidago sp* than to incorrectly list *Solidago canadensis*.

lunate (loon' ate) – Crescent shaped, like a new moon.

lyrate (lie' rate)- With an outline like a lyre, with a fat round tip, and smaller lobes along the sides, like a mustard leaf, *Brassica*.

M

macro- (mack' row) – 1) Large or long.
2) Botanists' swear word, Holy Macro!

maculate (mack you late) – Spotted, blotchy.

male - A plant or flower with only male organs (stamens), lacking female organs. Synonym: staminate. See monoecious, dioecious.

many – Numerous, usually describes the number of stamens in a flower, technically means eleven or more.

margin (mar' gin) – The outer edge, as on a leaf.

materia medica (ma tier' ee ah med' ic ah) – 1) Book dealing with the plants used for medicines.
2) The medicinal plants themselves. See herb.

matinal (mat' in al) – Bursting into bloom in early morning.

mealy (me' lee) – Feeling like a surface was dusted with fine corn meal. Synonym: farinaceous.

medial (me' dee al) - Along the lengthwise line (axis).

medicinal plants - 1) Herbs used for treating illness.
2) Plants which relate to physical symptoms, such as separating *Viola* often produces headaches; comparison of *Amelanchier* brings forth melancholy; and searching out alpine plants causes joint problems.

membranous (mem' brain us)- Thin, soft, and almost transparent.

Mendelian law (men day' lee an law) – The natural genetic inheritance of plants (or animals) with dominant and recessive genes. The monk, Mendel, did the studies on peas, *Lathyrus spp.*, with green and yellow color, and smooth and wrinkled skins. Three out of four of the seeds showed the appearance (phenotype) of the dominant gene, though the fourth seed and two of the middle ones carried the inheritance (genotype) for the recessive trait. See phenotype, genotype.

meristem (mare' is stem) – Group of cells that grow new tissues. The growing points of a plant. **Apical meristems** are those at the tip (apex) of branches and roots. **Secondary meristems** produce wood or bark.

-merous (mare' us) – A combining form used to show the number of parts or divisions; if a flower has 5 petals, 5 sepals, 5 stamens, it is 5-merous. See -cyclic.

mesic (mess' ick, mez' ick, me' sick) – A habitat with a well balanced water supply. See xeric, swamp, marsh.

mesocarp (mez' oh carp) – The middle portion of the fruit wall; in a cherry, Prunus, it is the meat. See ecto- for full description.

mesophytic (mezo' fit' tick) – In reference to a plant which prefers average conditions, especially moisture.

microclimate (mike'row climb' it) –Small areas where the temperature and moisture differ from its surroundings; like a frost pocket or hillside.

microenvironment (mike' row en vire'n ment) – A small area different from its surroundings, perhaps where rocks abound, or under a tree.

midrib - The biggest vein that runs down the middle of a leaf.

milk – 1) A white thick plant juice (sap) as from milkweed, *Asclepias*.
2) Immature seeds are still being fed and are said to be 'in the milk'. They need to be left to harden before being collected.

minimum - A very small member of the Asteraceae (Composite) family.

minute (my noot')- Very small.

modified leaf - Ancient plants developed leaves. Over the eons, leaves have further changed to become petals, tendrils, suction cups, bristles, spathes, etc. One leaf modification is the fruiting bract of the basswood, *Tilia americana*.

monadelphous (ma' ah dell' fuss) – With male parts (stamens) having their filaments scrunched together into a single group.

monandrous (moan and' russ) – With one lonesome male part (stamen).

mono- (ma' no) - A prefix meaning one, as monochromatic defines species whose flowers come in only one color, though that color may vary in shades and tints like red and pink.

monocarpic (ma' no car' pick) – In reference to a plant that lives for many years, blooms once to set seed, and then dies.

monoclinous (ma no kline' us) - With perfect flowers, that is, each flower contains male (stamen) and female (pistil) parts. The clin- part comes from a word meaning couch, and mono- means one, so the male and female parts are on one couch, or in the same flower. See monoecious and dioecious for comparisons.

monocots (ma' no cots)- Shorthand for the monocotyledonous plants, with one seed leaf, parallel veins in the leaves, and usually three petals (or multiples of three) in the flowers. There are a few exceptions.

monoculture – A human invention where a single species of plant is grown without any other species intermingled. Hundreds of acres of corn, Christmas tree plantations, and extensive fields of tulips are examples. It is prone to severe problems such as easy access of pests and plant diseases. It demands constant monitoring or obnoxious amounts of chemicals. See IPM, organic gardening/ farming, rotation.

monoecious (ma knee' shus)- Referring to a plant that has some flowers with male organs (stamens) and some flowers with female organs (carpels, pistils), but rarely flowers with both male and female organs (perfect). The oec- part comes from the same base as ecology (house), with mono- meaning one, so the blossoms are present in one house.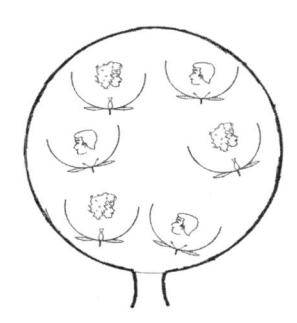

montane (mont ain') – Grows in the mountains. Synonym: alpine.

morphology (more foll' oh jee)- The study of shapes as botanists do with plant parts in the herbarium, and teens do from a street corner.

mother plant – A mature, unattractive older plant that is used to produce slips or cuttings for reproduction.

mouth – The open end of a bell or tube flower.

mucilaginous (mew' sill aj' in us) - Gummy, like glue leaking out.

mummify (mum' if eye) – When a fruit does not develop properly (aborts), the fruit turns hard or papery with no seeds.

muricate (myoor' ih kate) – With small warty bumps.

mutation (mew' tay shun) – An unexpected change in the genes of a plant, leading to a new character, such as a plant with blue flowers producing seeds of a plant with red flowers. Mutated characters are generally of little value, but occasionally produce good results. Mutations are passed to the next generation (inheritable).

mycelium (my seal' ee um) – The persistent hidden branches of a fungus, traveling outward underground from a center, surviving until conditions are right and producing a circle of organisms, like a fairy ring.

mycologist (my coll' oh jist) – A person who studies mushrooms and toadstools (fungi). Since the two are hard to separate, and toadstools are poisonous, there are old mycologists and bold mycologists, but no old, bold mycologists.

mycorrhizal (my' core rise' al) – Needing a certain fungus around the roots to grow well.

N

nacreous (knack' ree us) – With a surface color like the inside of a clam.

naked - 1) A bud without scales.
2) Any unprotected part.

napiform (nap' ih form) – Turnip shaped, said of roots.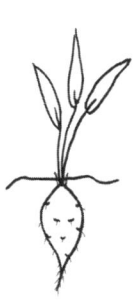

natant (nay' tant) – Found in water.

native plant – One already adapted to the local temperature and moisture (climate) as well as plant pests and diseases. Antonyms: alien, exotic, foreign. See naturalized.

naturalist – A biologist who explains about plants and animals, including reproduction, to the general public. Not to be confused with a naturist, a person with no clothes who may be inclined to demonstrate his talks.

natural balance – Nature takes care of most problems if allowed to. Poisons and introduced control agents often cause more problems than they solve. Restoration of local pest enemies is nearly always safe. One of the classic negative examples of introduced aliens occurred in the Caribbean Islands in 1877. Sugar cane workers feared the Norway rats, so the mongoose was brought in. The competing tree rats evaded the non-climbing mongoose. Hungry mongooses (mongeese?) ate snakes that had controlled the tree rats who took over the Norway rat territory. The mongooses multiplied to pest status, eating chickens and ducks.

natural controls – Such things as ladybugs, lace wing flies, preying mantis and toads brought in to control aphids and other insect pests. Chances are that these guys in white hats lived in the garden before people started using poisons instead of enduring a short outbreak.

naturalized plant (nat' your al ized plant) – An alien that fits the new habitat so well that it grows and reproduces as though it had always been there (native).

nectar (neck' ter) - Dilute honey produced by glands (nectaries) in the flower, a biological bribe by which flowers attract pollinators.

nectary (neck' tah ree) - The honey gland, appearing as a lump, a pit or a scale, usually found on the petals or in the spur of a flower.

needle - A modified leaf shaped like a sewing needle, as in a pine tree, *Pinus*, or cactus, *Mammalaria*. A thorn has a woody center, while a prickle is made of the 'bark' of the stem; neither thorns nor prickles are modified leaves. See thorn, prickle.

nerve - As soon as you figure out what a rib is on a leaf, botanists start calling it a vein. When you understand that, they switch again and call it a nerve. To be accurate, a rib is a support structure, a vein carries plant fluids, and a nerve is a ridge or fiber; to be practical, all three functions are usually present in the same structure.

net veins – Those interlocking veins found in dicots. See reticulate.

nettle - A highly developed plant, *Urtica*, which uses miniature hypodermic needles filled with irritants to successfully protect itself against more or less hairless primates, one of nature's jokes.
To apologize, nature usually places burdock, *Arctium*, nearby and the foot long leaves, crushed and rubbed into the welts, neutralize the burning itch.

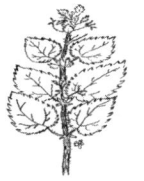

Others of nature's jokes include, but are not limited to, steep downhill slopes of wet clay, patches of *Smilax*, snakes, mosquitoes and no-see-ums, friendly dogs carrying poison ivy oil on their fur, spider webs across a path, and the occasional earthquake.

neutral (new' trull) - 1) Without sex organs, as some flowers and fixed dogs.
2) Soil with a pH of 7, neither acid nor alkaline.

new wood – Branches that contain growth material (meristem). Nurserymen starting new plants often prefer this type. Synonym: green cuttings, softwood cuttings. Compare with old wood.

nitrogen cycle – Nitrogen gas is taken from the air and fixed in soil as organic compounds. Plants pick up the nitrogen and when the plants no longer need it, the nitrogen is again released into the air.

nocturnal (knock turn' al) – 1) With flowers that open at night.
2) What you must become in order to pass the botany final exam.

nodding – 1) Drooping, hanging down.
2) What your head starts when you are nocturnal.

node - 1) A fat joint between sections of the stem, usually with one or more leaves growing from it.
2) Part of the human anatomy involved in allergy to pollen.

notate (no' tate) – Marked with spots or lines.

notch - 1) A little indentation in the outline of a part, as a notch at the top of a united petal, showing where the petals grew together.
2) The tip of an emarginate or retuse leaf.

notes - All persons interested in botany field trips should become proficient in writing notes. One of the first you should learn: "I just smashed into your parked car. Witnesses think I am leaving my name, address and insurance company. They are wrong."

nut - 1) A one-seeded dry fruit with a woody outer wall (pericarp) that does not break open at maturity (indehiscent).
2) A dry fruit who never breaks loose from botany to have other kinds of fun.

O

ob- (to rhyme with cob)- 1) Some part that got obstinate and grew backwards, for example, an obcordate leaf is heart shaped, but the fat part is close to the tip instead of close to the normal stem end. Upside down, turned over, or inside out.
2) A handy way to keep track of your backward blind dates: obFred and obMargaret.

oblanceolate – Shaped like the point of a spear, but with the wide part farthest from the attachment. See lanceolate.

obligate (ob' lig ate) – So obsessive that it will not consider any other lifestyle. Cannot survive unless conditions are exactly right. Examples: Woodland wildflowers will die without shade, sunflowers will die in dense shade.

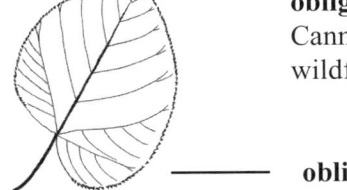

oblique (ob leek') – A base with sides that are <u>not</u> symmetrical, are of different shapes. Synonyms: unequal, asymmetrical, slanting.

oblong (ob' long) –With nearly parallel sides, two to four times longer than it is wide. Wider than **linear**, but not as wide as **square**.

obovate – Shaped like the outline of an egg, but with the wide part farthest from the attachment. See ovate, lanceolate.

obscure (obs cure') - Hard to find, hidden.

obsolete (ob' sole lete) - No longer useful, absent from a place where you would expect it to be, as obsolete petals on a wind pollinated flower. Barely visible, rudimentary.

obtuse (ob toos')- Blunt or rounded at the tip (apex) instead of pointed.

obvolute (ob' vole lute) – A pattern of leaves in bud (vernation) where each leaf overlaps the other in such a way that one half is exposed and the other half is covered.

Occam's Razor – Cut to the simplest explanation, usually right, e.g., which came first, vine or tree? No need for vine until it has something to climb, so the tree probably came first.

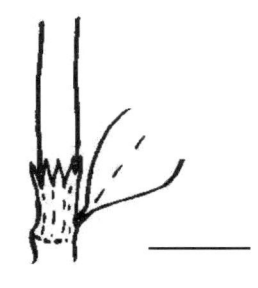

ocreate (ock ree ate) – In reference to a leaf attachment (petiole) growing from a node with an unusually large outgrowth (stipule) that spreads to surround the stem. Most common in Polygonaceae.

odd-pinnate - A compound leaf with the leaflets opposite each other and one leaflet without a mate at the end of the stalk. See pinnate for illustration. See appendix – leaf - arrangement.

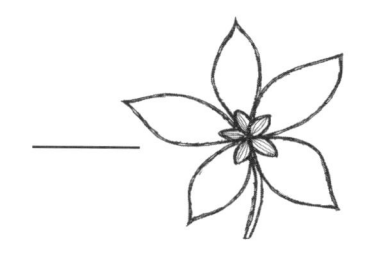

old wood – Vegetative cuttings taken now from last year's growth for propagation. These may be preferable for cuttings that need to be stored for longer periods, perhaps buried in damp sand until callus forms. Synonym: hard wood cuttings. Compare with new wood.

opportunist (opp' pore tune' ist)- One that takes advantage of the right conditions. A flower that normally blooms in spring may wait until summer if spring is too dry, giving an additional chance to set seed.

opposite – 1) Facing something, as an opposite sepal is located right in the mid-back of the petal, as shown on the back of this flower.
2) Leaves that grow directly across from each other on the stem, with two leaves per node.
3) Matched pairs.

optimal – The best result for the lowest cost in time, energy or money.

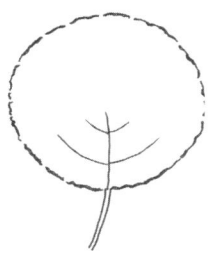

orbicular (or bick' you lar)-
1) Perfectly circular.
2) Located in a circle.

organic gardening/farming – Growing plants while using the smallest amounts of commercial chemicals, like fertilizers, herbicides, and pesticides.

organs - Plants as well as animals have organs. Organs are made of a group of tissues, and tissues are made of cells that are similar. A group of organs make up a **system**, and a collection of systems forms the **organism**. Think of a city as an organism. In this city/organism, **cells** are the bricks of a house. A bunch of bricks form a wall, the **tissue**. A bunch of walls form the kitchen, a room, similar to an **organ** that performs a certain function. All of the rooms (organs) form a **system**, the house. Within the city (organism), you find other systems, such as the water treatment plant (distribution of water), hospital (repair), trucking company (distribution of food, lumber, etc) and garbage company (disposal of wastes). The flowering plant is the organism. It has systems for reproduction, food production, water collection and waste disposal. Each of these systems has organs, such as leaves, roots, etc., and those organs are made of tissues, which in turn are composed of cells.

ornamental – (The word is normally an adjective, describing something; here it is used as a noun, the name of something.) Many ornamental trees serve only the one purpose of lovely spring flowers, beauty. Several plants with beautiful flowers may be available that also provide food and shelter for wildlife.

ornithophilous (or nith aw' fill us) – Pollinated by birds.

ovary - The middle of the flower where the eggs (ova) are produced, that grows the seeds after fertilization (pollination). Synonym: ovulary.

ovate – A shape resembling the outline of an egg, with the wide part closest to the attachment.

ovoid - A three dimensional figure like an egg.

ovule, ovum, pl. **ova** (oh' vyule, oh' vum, oh' vah)- The female reproductive part (egg) of a flower that matures into a seed.

P

painted – Colored with unequal streaks on another color, like a painted trillium, *Trillium undulatum*.

palaceous (pal lace' ee us, pal lah' see us)- With a papery cover.

palatable (pal' at ah bull) – Good tasting but not always healthy.

palate (pal' ut)- The 'tonsils' in the throat of a two-lipped (bilabiate) flower.

paleobotany (pail' ee oh bot' annie) - The study of plants that dinosaurs ate and stomped on, by finding their remains in rocks as fossils.

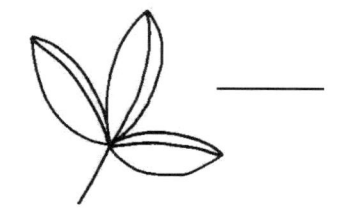

palmate (palm' ate)- Spread like the fingers of a hand, usually said of a compound leaf, but also refers to the veining of some simple leaves such as maple, *Acer*.

palustrine (pal' us tree n) – In reference to an area that is not part of a lake being subjected to waves and currents, but constantly wet all year; a marsh.

pandurate (pan' dur at)- Round at both ends and squeezed in at the middle, like a violin or hourglass.

panicle (pan' ih cull)- An indeterminate branching flower cluster (inflorescence) with the individual flowers on little stalks (pedicels), like lilac, *Syringa vulgaris*.

pannose (pan' ohs) – Covered with felted hairs.

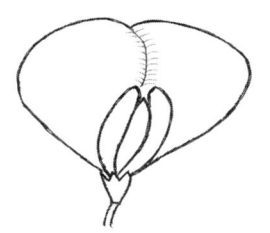

papilionaceous (pap ill' ee on nase' ee us, pap pe' yon nase' ee us)-
Like a butterfly, said of flowers with wing-like petals, such as the sweet pea, *Lathyrus*.

papilla (pap ill' ah)- A pimple, mostly observed on teen-aged flowers.

pappus (pap' us)- The papery part of a seed head that isn't seeds, growing between the seeds, as if the plant had put it there to confuse you as to where the seeds are hiding. In the illustration, seeds are black, pappus is light. Each is attached to a receptacle, the left flower a cylinder, the right a mound.

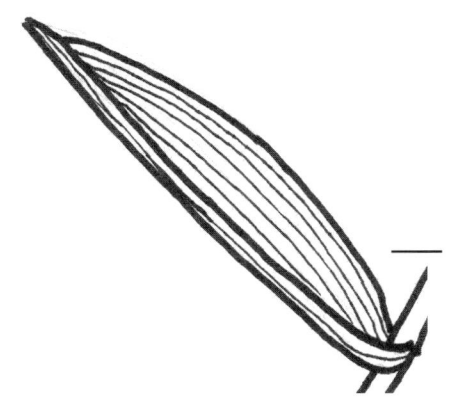

parallel (pair' all ell) - Side by side, for example, parallel anthers enclose the stamen stalk (filament) like the two parts of a bun around a wiener.

parallel veins - Those ridges that run from the base of a leaf to the tip, running side by side for the whole length, as in grass leaves, Poaceae (Graminae), looking like the ridges of corduroy cloth.

parasite - 1) A plant drawing its food from other living plants, like mistletoe, *Phoradendron*, or dodder, *Cuscuta*. A heavy infestation may kill the host plant, but usually doesn't. See saprophyte.
2) Anything which takes a lot out of the host and gives little or nothing in return, e.g., thick botany books with no pictures, and neophyte students who don't bring goodies to a study session. See host.

parted - Cut or cleft almost to the base, half-way to three quarters of the distance from margin to midrib, or tip (apex) to bottom (base).

pasturing – Labor saving practice of having animals harvest the crop directly from the field by eating it, saving time and energy for the farmer.

path - The long way around which is the shortest distance between two impassable habitats.

pathogen (path' oh jen) – Any micro-organism that causes disease.

pathology (path oll' oh gee) – The branch of science that deals with the study of all aspects of disease, especially after death.

pectinate (peck' tin ate) - Like a comb; with long, stiff hairs growing very closely together, often found on obFred and obMatt.

pedicel (ped' ih sell)- The individual flower stalk in the midst of a cluster of flowers (inflorescence), a pedestal where a single flower rests. See peduncle.

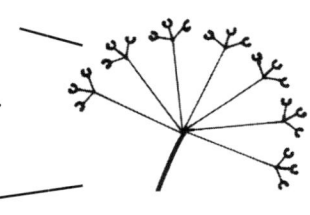

peduncle (peh dunk' ull) - The main stalk for a complete group of flowers (infloresence) or for a single large flower.

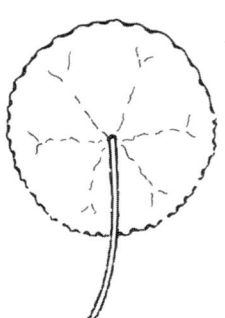

peltate (pell' tate)- Refers to a rounded leaf with the stalk attached in the middle rather than at the side, rather like an umbrella, found in Mayapples.

About fourteen hundred Mayapple stems,
With their parasols up, marched down the hill
And all the Spring Beauties turned up their pale, peaked
noses
And said, "Don't them Mayapples
Think they're somebody
With their bumbershoots up!"
—*Jake Falstaff*

pendent (pen' dent) - Any hanging part.

pendulous (pen' dew luss) - Hanging loose or free, swinging like a pendulum.

pepo (pea' poh)- A large, thick-walled, leathery fruit (berry), like a gourd or pumpkin, *Curcurbita*, which—believe it or not—are botanical fruits. See fruit.

perennial - A plant that lives at least three years, or would have if it had made it through the winter after you took it home and planted it in the wrong habitat.

perfect - With a combination of male parts (stamens) and female parts (pistils) in the same flower, making it fertile.

perfoliate (per foal' ee ate)- Referring to plants on which the stem pierces through the middle of the leaves.

perforate (per' for rate) - With holes like windows in a leaf or petal, that were there by nature, not left there by a hungry bug.

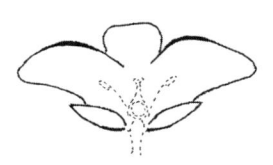

perianth (pear' ee anth)- All the sepals and petals. The fancy packaging around the working parts of a flower (pistil and stamens). The petals (corolla) may be called the inner perianth, and the sepals (calyx) the outer perianth. Synonyms: floral envelope, accessory organs.

pericarp (pair' ih carp) – The outer layer of the fruit wall, the peeling. See ecto-.

peripheral (purr if' er al) - On the outer edge, where you may be driven after trying to learn all these terms.

perpendicular (purr pen dick' you lar) - Sticking out at a right angle to another line (axis). See horizontal.

persistent (purr sis' tent) - 1) Hanging on, staying attached; said of individual parts like the sepals (calyx) on the bottom of an apple, *Malus*.
2) In reference to juvenile features that normally disappear, like hairs or reddish color.
3) What you must be to pass botany the first time around.

pest - 1) An insect or disease that harms plants, animals or people. See natural balance.
2) An insect such as a gnat, or snake, or farm dog, or other natural bother for humans. See natural controls.
3) An animal or plant that competes with man. It is fine for a coyote to eat rabbits and rats, but when they eat chickens, they gotta go! Same with weeds in the garden, though many of the weeds are more nutritious than the planted varieties.
4) Any fellow student, leader or prof who keeps asking you questions for which you don't have the answers.

pest control - 1) Insect repellant.
2) Telling a pest about a shortcut for the recommended path. See path.

pest resurgence – An increase in the number of pests after use of broad spectrum pesticide, usually because it destroyed natural enemies that had kept the pest in check.

petal - The individual panel of fancy wrapping of a flower that helps to attract the birds and bees who make reproduction happen. Part of the corolla, therefore a usually colorful leaflike part of the perianth.

petaloid (pet' al oid) - Refers to something that looks like a petal, though it functions like some other part, such as a stamen; both pretty and functional.

petiolar (pet' ee oh lar) - Referring to or located on a leaf stalk (petiole).

petiolate (pet' ee oh late) - With a leaf stalk (petiole). Synonym: stalked. Antonym: sessile.

petiole (pet' ee ohl)- A leaf stalk.

pH (pea aitch) – The measure of acid (hydrogen ions) in a compound. Neutral is 7, with lower numbers being more acid, those higher are less acid, more alkaline. Logarithm is used in the numbering so each rise of one number is a ten times change in acidity. Rising two numbers is 10 X 10 or a hundred times as great. Let's say a lake should be neutral, pH of 7. Various factors make the water more acid, dropping the pH to 5, 100 times as much hydrogen as it held at 7. This pH does not allow the calcium in the water to stay in solution so it turns to a solid (precipitates out) called marl which sinks to the bottom. Few plants can survive at pH of 5, mostly carnivorous ones. See carnivorous.

phenology (fee noll' oh gee) – 1) The branch of science studying the relationships between plants and climate, such as when a plant will flower in different zones or altitudes.
2) Not to be confused with phrenology, which is studying bumps on your skull to determine your mental faculties and character, but were more likely caused by you being a legend, or from you turning your field specimen collection in late. See legend.

phenotype (fee' no type) – The appearance of a plant due to its genetic makeup (genotype). Two plants with the same appearance (phenotype) may have different genotypes, as a botany professor found out the hard way. He hired students to collect seeds from white and pink bluebonnets, *Lupinus spp.* These were carefully separated and sold at high prices in packets with brilliant colors. Sadly for the prof, the seeds followed Mendelian laws of heredity, and the genotypes produced 75% of the flowers with the phenotype of blue color. See Mendelian law, genotype.

phloem (flow' em) - 1) Vessels similar to mammalian arteries that carry nutrients to the upper parts of the plant.
2) Phirm philamental phlumes, phurnishing phast phood phor plants.

photosynthesis (foe' toe sin' the sis) 1) The process whereby green plants convert nutrients into food by using sunlight.
2) The process used by photographers to produce perfect illustrations that look nothing like the bug-chewed, wind-blown, student-stomped specimen you find in the field.

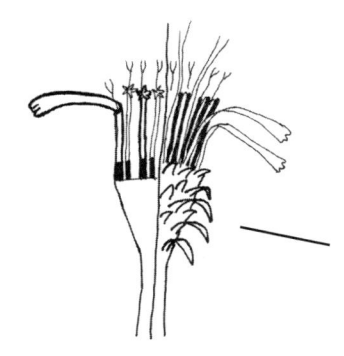

phyllary (fill' ah ree)- One of the little leaflets on the bottom (involucre) of the flower head of daisy types, Asteraceae (Compositae).

physic (fizz' ick) – 1) ancient word for medicinal plants, from which the modern word 'physician' comes. Early botanical gardens at medical schools were often called Physic Gardens.
2) Laxative, which many ancient medicines were.

physiology (fizz' ee ol' oh gee) - The branch of the biology that deals with how internal organs and processes function.

phytotoxic (fight oh tox' ick) – A ten dollar word to use when your neighbor hires a chemical company to care for their lawn and the spray drifts over to your flowerbeds with deadly results.

picot (pea' ko) – With a narrow edging of contrasting color, usually white, on a flower petal.

pilose (pie' lohs)- Lightly covered with straight silky hairs.

pinching - 1) A process of removing terminal (apical) leaves to increase new branch growth, changing of hormone production. See pruning.
2) What students of the opposite sex may attempt when you are in the woods, hoping for an increase of hormone production. See pest control.

 **pinnate** (pin' ate)- Like a feather, said of the pattern of veins in some leaves or of the leaflets of a compound leaf.

pioneer plants – Hardy plants that can survive under adverse conditions. They move in after fire, clearcutting, landslides, eruptions and bulldozing, their roots stabilizing and loosening the soil for more desirable plants to grow. Synonym: weeds.

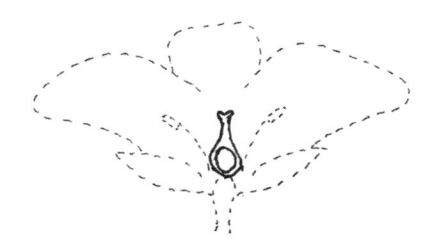 **pistil** (pis' till) - All the girly parts of a single simple flower. It may contain multiple carpels in a compound flower. Synonym: gynoecium.

pistillate (pis' till ate) - Said of flowers with only female organs (pistils) and no male organs (stamens).

pitcher - 1) A leaf that has been modified into a tube, often largely swollen on one side, to capture insects, as in pitcher plant, *Sarracenia*.
2) The order you place at the first bar after returning from a field trip.

pith - The tissue inside a stem, usually soft, but may be solid. To find the type of pith, a stem must be cut open lengthwise. Types of pith are **chambered**, **continuous**, **diaphragmed**, **hollow** and **spongy**.

plant patent – People work hard at producing a new color, shape or size of plant, most often referring to color. To protect their investment, they apply for a plant patent so others cannot use their work to make money. It is illegal to make cuttings of patented plants.

pleated - Folded lengthwise several times, common in buds holding embryo petals and leaves. See plicate.

plicate (ply' kate) – Folded like a pleated skirt. Pleated, folded lengthwise. This pattern is found in buds of united corollas, with either sepals or petals as shown in this cross section. Think of it as an umbrella folded for storage.

-pli-nerved - A suffix used with a number, describing a leaf with three or more large veins (ribs), the side (lateral) ones growing out from above the base of the middle one, as '3-pli-nerved' or '3-nerved' would have two side veins plus the middle one. See nerve.

pod - 1) Fruit of a legume that splits (dehisces) down two sides, such as a pea or bean, Leguminosae.
2) The term is often used incorrectly for any fruit with a papery covering.

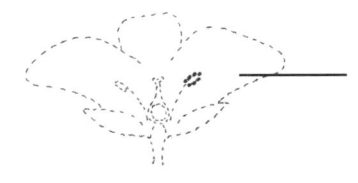

pollen - 1) The powdery, grainy or waxy parts of a flower which contain the male cells (sperm).
2) What gets on your chin from a dandelion when someone checks to see if you like butter.

pollination - The act of fertilization in plants, most commonly by insects carrying pollen from the stamen to a pistil, but may travel by wind or water. Nature has many tricky types of pollination, such as having the male part (stamen) and female part (pistil) ripen at different times to insure out-breeding to keep a species strong. At the opposite extreme stand the violets, *Viola*, that produce lovely barren flowers in the spring, but then produce closed (cleistogamous) flowers that following summer.

pollinators – Insects and other animals that act as go-betweens to carry pollen from male flowers to female flowers; mobile sperm banks.

pome (to rhyme with home) - A fleshy fruit, like apple, *Malus*.

population density – The number of plants per unit, as planting 1000 Christmas trees per acre.

pore - 1) A small round hole.
2) What the clouds commence to do on the day you planned a picnic following the field trip.

posterior (poss tee' ree or)- 1) Closer to the stem (axis), on the back side of the flower.
2) Where you are tempted to kick the student who lets his specimens wilt, then wants you to share yours.

prickle - A spine that comes from the stem covering or bark, sometimes from a modified leaf, rather than from the wood. Rose "thorns" are actually prickles. Other terms for prickles are: spine, tine, spiculum, needle, pin, pricker, spur, rowel, barb, cusp, snag, bristle, beard, brier, briar, bramble, thistle, comb, awn, glochid and bur. See thorn, modified leaf.

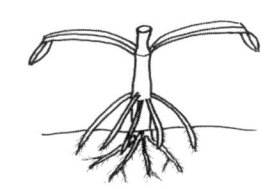

procumbent (pro come' bent)- Trailing, or lying flat, but not rooting at the nodes.

profile – A vertical section of soil all the way from topsoil to bedrock.

propagation (prop' ah gay' shun) – The many different ways of bring new plants into your life.

prostrate (pra' [to rhyme with bra] straight) – Spread out flat, and may root at nodes. Includes creeping, decumbent, procumbent and trailing, but not dead.

prop roots – Roots arising from the above-ground stem (adventitious roots) and spreading outward to produce stabilizing balance for a tall plant like corn, *Zea mays*.

protected bud - A bud surrounded by scales. See bud and naked.

proximal (prock' sim al) - Close to the point of attachment. Antonym: distal.

pruning (prue' ning) – Cutting off leaves or branches for many different reasons such as shaping, removing dead or diseased parts, increasing yield of flowers or fruits, etc. This messes with the hormone production so be sure not to prune as autumn nears since it may take too much strength from the plant to increase foliage that will be killed by frost.

Pteridophyte (tare rid' oh fight) – A flowerless plant from the great group Pteridophyta, that includes ferns and other plants that reproduce with spores rather than seeds.

puberulent (pew bur' you lent) - With barely visible hairs, like a pubescent boy's first whisker.

pubescent (pew bess'sent)- Technically speaking, softly hairy, like a boy's mustache at puberty. In general use, it includes any type of hairs, just as boys' mustaches vary as they develop.

punctate (punk' tate)- With tiny windows, translucent or colored dots, dimples or pits.

pustular, pustulate (pus' two lar; pus' two late)- With blisters, looking like infected cuts with pus.

pyriform (pie' ree form)- Shaped like a pear, fruit of *Pyrus spp*.

Q

quadrate (quad' rate) – Square, like the stem of a mint, *Mentha spp*.

quadri- (quad' ree) – Four of something.

quinque- (kwin' kway) – Five of something.

R

raceme (ray seem') - A long conical cluster of flowers (inflorescence) with short stalks (pedicels), each of which grows directly from the main stem. Flowering starts at the bottom and the main stalk continues to grow (indeterminate), forming new buds at the tip (apex). The main stalk is called a peduncle below the flowers, and a rachis inside the cluster.

rachis (ray' kiss) - The center rib of a compound leaf above the petiole.

radial (ray' dee al)- 1) Said of a flower with all the parts extending equally from the center like spokes from a wheel. Synonyms: actinomorphic, radiosymmetrical.
2) A type of tire which gets a workout during field specimen collections.

radical (rad' ih kal) - Arising from the roots; said of basal rosettes. Nothing to do with politics.

radicle (rad' ih kul) - The embryo root in a seed.

raising the crown – Removing lower limbs on a tree. See high shade.

ramose, rameal (ray' mose, ram' ose; ray' me al) - Referring to a plant that is branched.

range - 1) The borders of the geographical area where a plant may be found, that may be political like a state; surface feature like a mountain range, or dealing with location like an ocean shoreline.
2) Where cows find a home, and deer and antelope play.
3) A stove for cooking cow, deer and antelope, along with edible plants.

rank - 1) Level of classification (taxa), including but not limited to: kingdom, division, class, order, family, genus and species.
2) How your housemates describe your odor when you return from a collection trip.

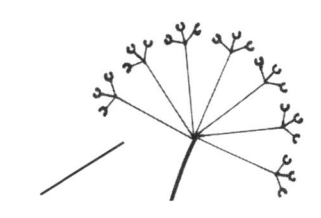

ray - 1) A tube flower in which the petals have migrated to one side, looking like a petal on the head. Synonyms: strap or ligulate florets.
2) The medium stalks in between the main flower stalk (peduncle) and the tiny stalk of the individual flower (pedicel) that forms the circular shape of the group of flowers (inflorescence) in a compound umbel, as in queen anne's lace, *Daucus*.

receptacle (ree sep' tah cull) - The enlarged end of the flower stalk (peduncle) which holds some or all of the flower parts, very apparent in the daisy family, Asteraceae (Compositae), where the receptacle holds the florets. Receptacles come in many shapes, but the flower parts and eventually the seeds (shown in black) are attached, here to a cylinder and a mound.

recessive (ree sess' ive) – In reference to an inheritable trait (gene) that is likely to be overpowered by a dominant trait. Usually expressed by a lower case letter (r) while the dominant trait is in upper case letters (R). Typical Mendelian pattern would be RR, Rr, Rr, rr, where dominant trait R would show (phenotype) in 3 out of 4, even though the recessive trait r is also present in 3 (genotype).

reclined (re kline'd) - 1) Bent downward. See descending.
2) The position taken while catnapping during lecture.

recumbent (ree come' bent) - Leaning, lying flat on the ground. See decumbent.

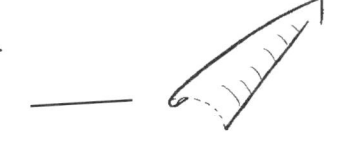

recurved (ree curved') - 1) Curled backward, as the tip of a leaf curving toward the attachment, somewhat hiding the under side of the leaf.
2) Similar in shape to how your tongue looks when the doctor tells you to say, "AHHH". Opposite of incurved, which is like your tongue tip reaching for your nose tip.

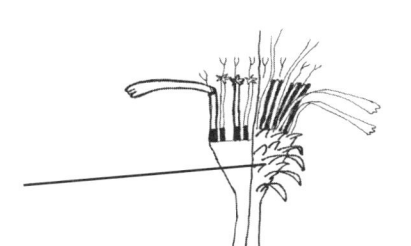

reflexed (ree flecks'd) - Like recurved, but a sharper angle, more like your elbow when you slap the mosquito biting your cheek.

regular - 1) Symmetrical, as in a flower with petals of the same size and shape.
2) Having similar things repeated; one of the crowd, not individualistic.

remote - 1) Spaced far apart.
2) What your mate accuses you of being while you are filling class requirements.
3) What you give your mate to keep busy using while you finish class requirements.

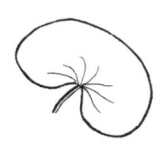

reniform (reen' if form) - Kidney shaped; circular or oval with one side dented in.

reproductive organs – The male (stamen) and female (pistil) parts necessary for the plants to form seeds. Synonym: essential organs. Antonym: accessory organs.

resinous (rez' in us)- Sticky.

reticulate (re tick' you lat) - With veins arranged like a net; not parallel veined, often easier to see on the back of the leaf. Found mostly in dicots.

retrorse (reet' roar s) – Bent backward or down.

retuse (reh toose') - With a shallow notch on a rounded tip.

revolute (rev' vole lute) - Having side margins that roll downward around the bottom surface. See involute, recurve.

rhizome (rye' zome) – A fat horizontal stem pretending to be a root, often acting as a storage organ. Rhizomes have nodes and/or buds, to help distinguish them from true roots that have neither. The nodes may produce new plants (asexual reproduction).

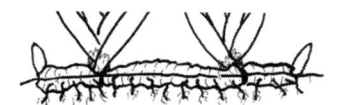

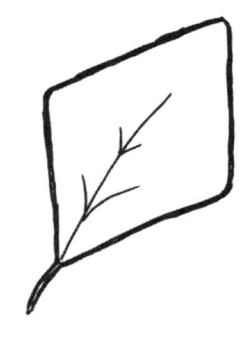

rhomboid – Diamond shaped, with the attachment at one of the narrow points.

rib – 1) The main vein, as down the center of a leaf.
2) Fibers in clusters forming corrugations on the surface, looking somewhat like corduroy cloth on most monocots. See nerve.

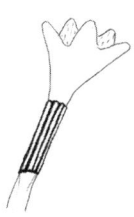

ribbed - 1) With a surface having veins or fibers running the long way (longitudinal), so the surface is delicately corrugated.
2) What you get after falling in the mud.

ringed - Striped or grooved crosswise on a cylinder like a stem. See striated, striped.

riparian (ripe pair' ee an) - Referring to the bank of a stream, as in "Creeping Charley, *Lysimachia nummalaria*, grows in riparian habitat."

rolled - 1) Having the side margins turning loosely, either up or down.
2) Referring to exciting actions that are very different in the bad section of town and in a private woodland glen.

root – The most common underground part of a plant which anchors the plant to the ground and soaks up food and water.

rosette (rose et') - A cluster of leaves all coming from the same spot, usually on ground level where it is called a **basal rosette**.

rostrate (rost' rate) – With the form of a beak.

rot - 1) Bacterial or fungal deterioration of plant tissue, as specimens left in the back seat of a car in the sun.
2) Supposedly helpful advice from a first year botany student.

rotate (row' tate) - Shaped like a wheel, said of a flower with a very short tube of united petals flaring into a flat top.
  (back view)

rotund (row ton'd)- Fat and round, said of such things as the corolla of some united petal flowers like a blueberry, *Vaccinium fuscatum*.

ruderal (rude' er al) – Growing in disturbed places, often pioneer plants.

rudimentary (rude ih ment' ary) –
1) Not completely developed, no longer needed, nonfunctional.
2) Present but not working, rather like botany students on the Monday after Homecoming.

rugose (rue' gohs, rue gohs') - With wrinkles, bumpy.

runner - 1) A little plant running away from home while still holding on to the mother plant by a long, slender stalk (stolon), then later taking root at the tip.

2) A botany student who stepped in a yellowjacket nest.

S

saccate (sack' ate) – Like a sack or pouch.

sagittate (saj' it ate)- Shaped like an outline of an arrowhead. The downward ends may be pointed or round. See hastate.

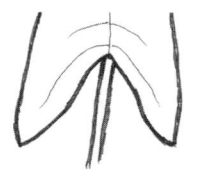

salverform (sal' ver form)- Referring to a flower with a long tube which flares out into a flat top. Compare with rotate, tubular.

samara (sah mare' ah) - 1) A winged dry fruit that does not break open at maturity (indehiscent). Found in maple, *Acer.*, and ash, *Fraxinus* and more.
2) Others of, as when you ask someone who has found a plant you need for your collection, "Do you know where there's samara those?"

sanguineous (san gwin' ee us) – Blood red.

sap - 1) The juices of a plant, similar in function to blood in people.
2) A human pest.
3) A clavate tool to use in protecting yourself from #2.

saponaceous (sap' on nase' ee us) – Soapy.

saprophyte (sap' row fight)- A plant living on dead organic matter, not a parasite, typically not green, such as Indian pipe, *Monotropa*.

saturated (sat' your ate ed) – 1) In reference to soil moisture, when the soil can't hold any more water. See water table.
2) How your brain feels just before a botany exam.

saxatile (sacks' ah teal) – Growing in rocks, especially in dry areas.

scabrous (scay' brus)- Scabby, with rough projections, not soft and smooth.

scale - A tiny modified leaf that is not green.

scandent (skan' dent)- Climbing. See vine.

scape (skay p)- A flower stalk without leaves (naked), though it may have scales or bracts. Being stalked naked is dangerous, and the flowers are often short-lived.

scarify (scare'if'eye) – To nick or scrape through a hard seed coat like morning glory, *Convolvulus*, to allow moisture to enter, to speed germination.

scarious (scare' ee us)- 1) Not green; papery.
2) Referring to the upcoming exam.

scarred - Having a surface with old marks showing where the leaf bases or other structures had been. Marks within the scar will identify species.

scattered - Irregularly arranged, as leaves not completely alternate or opposite. Separated, loose.

schizo- (skizz' oh, skit' zoh) - 1) A combining form meaning split, divided, separating.
2) A specific personality problem often found in overly perfectionistic botanists.

schizocarp (-carp) - 1) A dry fruit that opens (dehiscent), splitting into sections, each holding a single seed, with each piece looking like a simple fruit.
2) A fish with personality problems.

scientific name – Binomial, technical name.

scion (sigh' on) – Cutting from a desirable woody plant. See grafting.

scurfy - 1) With crusts of scales that rub off.
2) In reference to many student rooms.

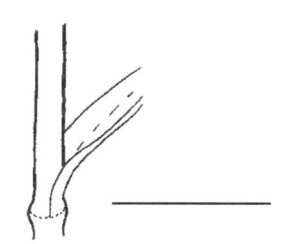

sebaceous (sah base' ee us) – Feels like hard fat.

secondary growth (seck' un dare ee) – Any growth which does not happen at the tips (meristem) of stems or roots, such as production of wood and bark in a tree. Growth that makes parts fatter, not longer.

secund (sek' und)- With all structures hanging from one side, like the boys at the bar. Synonym: unilateral.

seed - A mature, fertilized egg, ready to grow.

seed coat - Outer protective coating of a fertilized ripe egg.

seed tree – A mature tree of desired species left to live and reproduce in an area being clear cut (timbered). See wolf tree.

seedling – A tiny plant that has recently germinated.

segment - A part that is deeply divided, but not really compound, like a lobe on a petal, or an ear on a leaf.

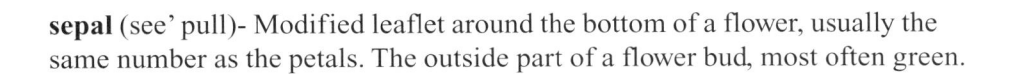

sepal (see' pull)- Modified leaflet around the bottom of a flower, usually the same number as the petals. The outside part of a flower bud, most often green.

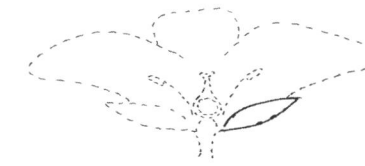

septicidal (sep' tea side' al) – A pattern of opening (dehiscense). See dehiscent.

septum (sep' tum) - A partition or cross-wall between two sections; the tissue that divides a human nose into two nostrils is a septum.

seriate (sear' ee at)- In a series, like the whorls of leaves up the stem of tiger lilies, *Lilium*.

sericeous (sair iss' ee us)- Nothing to do with series, just means with silky hairs.

serrate (sair' ate)- In reference to a margin like the teeth of a saw.

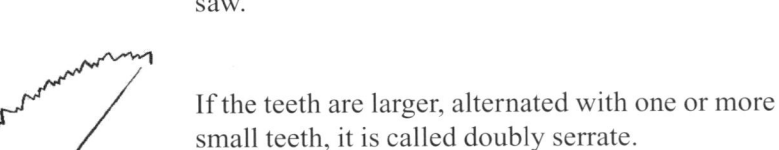

If the teeth are larger, alternated with one or more small teeth, it is called doubly serrate.

sessile (sess' ull)- Referring to a flower or leaf without a stalk, attached directly to the stem.

set – A vegetative plant part that will grow, like an onion set.

sheath - The tubular base of a leaf, partly surrounding the stem.

shining - With a surface that looks polished.

shoot - 1) A non-technical term referring to new growth from any part of the plant.
2) What you say when you open that pressed specimen and find it moldy.
3) What your instructor threatens to do to you if you don't get a new specimen to replace the moldy one.

shrub – A medium sized perennial woody (ligneous) plant with several stems.

silique (sill eek') – A dehiscent fruit with two openings, the seeds attached to a membrane rather than to the pod.

simple - Single, not compound, not branched, not divided, the stripped down model.

simple fruit - A fruit that grows from the ovary of a flower with a single female organ (carpel) in a single flower.

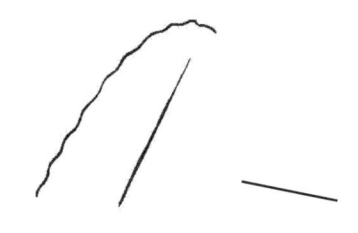

simple leaf - Plain old leaf with a stalk (petiole) and blade.

sinuate (sin' you ate) - Like a snake, curvy, with wavy margins.

sinus (sigh' nuss) - 1. The open space between two lobes or teeth.
2. The human organ that reacts to pollen by swelling so it is no longer open.

slip – A cutting, usually of a side branch of a mother plant.

snag – A non-technical term for a prickle, or sometimes a splinter from wood. See prickle.

snake - Shy animal group which has one or more species in nearly every habitat, mostly non-poisonous.
Rarely seen except by those who are terrified of them and try to sneak by. Whistle or sing, the louder and more off-key, the better. They also prefer not to see you and will slither off.

soft science - Any specialized field or discipline, like psychology, sociology, anthropology, or political science, where you can soft soap your way through, since the criteria are difficult to measure accurately, and opinion counts more than facts. See hard science.

softwood cuttings - Young branches that contain growth material (meristem). See new wood.

soil - Dirt that falls in one of three categories: sand, clay or loam. They are easy to tell apart. Sand always looks dry until you step on it and get wet feet. Clay either has cracks you can lose your notebook in, or clings to your shoes in huge globs. Loam has lots of humus, splashes up on your clothes and fingernails, and stains them.

solitary - Growing alone, not in a group. May be said of the plant, the flower or an organ within the flower.

sp., plural **spp**. (ess-pea, ess pea pea)- Species. When you are trying to identify a plant that belongs to a genus with a lot of species—such as goldenrod, *Solidago*—and aren't sure of the species, list it by the name of the genus followed by sp., as *Solidago sp*. The final identification may include things that you have to find by a microscope, such as beaks on the tiny seeds. Flowers and seeds rarely occur at the same time, so you have to wait to collect both. It is more professional to list sp. than to make a wrong guess.

spadix (spade' icks)- The central flower stalk inside a leafy hood (spathe) like the Jack in a Jack-in-the-pulpit, *Arisaema triphyllum*.

spathe (spayth) - A large modified leaf, often colored other than green, that encloses the true flowers arranged on a fleshy floral spike (spadix). Found in Araceae.

spatulate (spat' you lat)- Having a spoon-shaped outline, an oval with one end drawn into a long point (attenuate).

species (spee' she s) - 1) Plants that freely breed and vary little from each other; the basic unit of classification. As the species name alone may be found in other genera, a species is identified by genus and species in a two part name (binomial). The word is either singular, as in "I have identified this species," or plural, as in "The many species of goldenrods confuse me."
2) Using specie as a singular form may prove embarrassing since it refers to coins, so should be used only to refer to dollar plant, pennywort and the occasional quarterhorse.

specimen (spess' ih men) - A dried plant selected to show what that classification breakdown (species) should look like.

specimen plant – A tree chosen for beauty. See ornamental.

spent – In reference to a plant that is past its prime, non-productive.

Spermatophyte (spur mat' oh fight) – Any plant that produces seeds. Synonyms: seed-plant.

spiculum (spik' you lum) – Usually a fine fleshy point. See prickle.

spider - See snake. Whistling doesn't help as spiders are tone deaf.

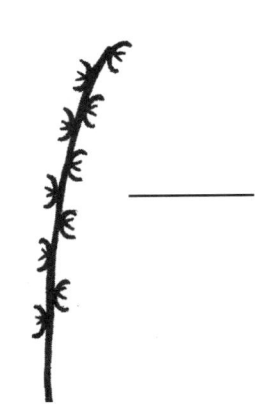

spike – 1) A straight-up flower stem, where the individual flowers grow directly from the stem <u>without</u> little stalks (sessile).
2) Loosely and incorrectly used for any long thin flower cluster (inflorescence).

spine - A rigid adaptation of a leaf, stem or other organ, usually modified to protect (arm) the a plant from being eaten or trampled. Cactus spines are most common, the leaf blade disappearing (evolving) to conserve water in a hot hostile environment. See prickle. See appendix - armor.

splitter – See lumpers and splitters.

spiral - 1) A part of a plant that twists like a corkscrew.
2) Used when leaves or flowers are arranged at regular intervals so they appear to be traveling around a central line (axis), moving upward or outward.

spongy pith - The soft, easily compressed material found inside some otherwise hollow stems.

spontaneous generation – The mistaken belief that microscopic plants can turn into animals and vice versa; believing that living things arise from dead material, like maggots come directly from rotten meat, and horsehair snakes appear from a hair from a horse's tail dropping into water and turning into a different animal.

sporangium (spore anj' gee um) - A case or sac where ferns hide their spores.

spore – Plants that don't flower cannot make seeds, so the spore is the reproductive body for plants like ferns and mosses.

sport – A new plant that shows up unexpectedly having a difference from its parents; a mutation. A good example is the navel orange, a bud-sport of the sweet orange, *Citrus sinensis*. Monks in Brazil discovered it around 1820, an orange with no seeds. Instead the ovary becomes a second rudimentary berry found within the orange where the blossom was—the "navel" in the orange. All navel orange trees grow from cuttings taken from that branch and its offspring. See mutation, berry, cutting, rudimentary.

spp. See sp.

spur – 1) A perky little tail on a flower, like columbine, *Aquilegia*.
2) A **fruiting spur**: short branchlet with closely placed nodes, that produces flowers and fruits.

ssp. (ess-ess-pea)- Subspecies. You have your work cut out if you are trying to identify clear down to subspecies. See lumpers and splitters.

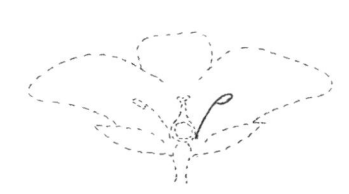

stalk - 1) A non-technical term that refers to the slender part attaching a small part to a larger support. See peduncle, stipe, petiole, scape, branch, etc. Any could be referred to as a stalk.
2) What borderline students tend to do to the instructor until grades are posted.

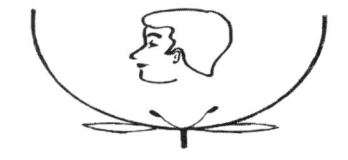

stamen (stay' men)- One complete male organ in a flower, usually made up of filament and anthers. The MEN part is a memory crutch that this is male.

staminate (stam' in ate) - With only male organs (stamens) in the flower. It is fertile, but needs contact with a female (pistil) to reproduce. The pollen from the stamen is carried to the female flower by different methods. See pollination.

standard - See banner.

stem – 1) The main support of a plant, the 'tree trunk' of a small green (herbaceous) plant.
2) Used loosely for any part that supports leaves or flowers.

sterile - Not fertile; not having male or female organs or other reproductive capabilities. Seen in plants such as horsetail, *Equisetum*, which has both sterile and fertile forms.

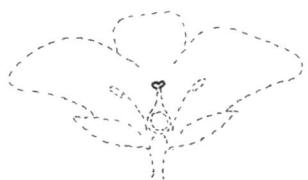

stigma (stig'ma)- The top of the female part (pistil) of the flower, sitting there waiting for the pollen. The MA part is a memory crutch that this is female.

stinging - 1) Having hairs that irritate the skin of botanists studying them. The hairs are usually erect.
2) Refers to unkind remarks made by those who think they know more than you do, which may or may not be true.

stipe (sty p) – The stem of a fern below where the leaflets are attached onto the rachis.

stipels (sty' pel s) - Paired scales, spines or glands at the base of the secondary leaf stalk (petiolule) of a compound leaf. See stipule, petiolule.

stipule (stip' yool)- A leaflet, scale, spine or gland that grows at the base of a primary leaf stalk (petiole), often in pairs. Some glossaries list stipel and stipule as synonyms.

stock – 1) The rooted part of a woody plant for grafting. See grafting.
2) The farmer's bull and his ladies who chase you across the pasture.
3) The shoulder portion of the farmer's gun if you rip through his fence while running from the cattle.

stolon (stow' lon)- A slender plant stem that grows sideways just above or under the ground, taking root and forming new plants. See rhizome and runner.

stoma; stomata (stow' ma; stow mat' ah) – A microscopic opening (pore), usually on the bottom side of a leaf, where the plant 'breathes', exchanging oxygen, carbon dioxide and water vapor.

stone – A common name for the woody pit (endocarp) of a drupe like a cherry, *Prunus*. See ecto-.

storage leaf - A succulent, fleshy leaf, especially hidden underground. See bulb.

stramineous (stra minnie' us) – The color of straw, yellowish.

strap flower - A floret in which all the petals joined to stick out one side, instead of forming a circle.
!If you look at the tip of the 'petal', you will see some notches. If there are four notches, this means five petals united in the strap.
Most strap flowers are around the outside of the head as in a daisy, *Chrysanthemum sp.*, with disk florets in the middle. In some, such as chicory, *Cichorium*, they are all strap flowers. Strap flowers may be fertile and form seeds, or they may be neutral (sterile).
Synonyms: ray floret or flower, ligulate floret or flower.

stratification (strat' if ih kay' shun) – 1) In nature, placing layers of leaves over top of seeds that fell in the autumn, rather like a blanket for a winter nap. The moist soil and overlay of leaves protect the seeds from being eaten. When spring comes, the seeds sprout and send shoots up.
2) People duplicate the process by placing seeds in plastic bags with a medium like sphagnum, putting them in the refrigerator or freezer, then planting them in the spring.

striate (stry' ate) - With lines running the long way on the surface. Synonym: ribbed.

striped - Having different colored lines running the long way of the leaf or flower. If the stripes are crosswise, the pattern is called banded.

strobilus; pl. strobili (stroh bye' lus; stroh bye' lye) – A tightly clustered bunch of wooden or papery leaf-like structures fastened to a central stalk, like a pine cone, *Pinus spp*, or hops flower, *Humulus*.

style - The neck of the female part (pistil), connecting the pollen collector (stigma) and the egg producer (ovary).

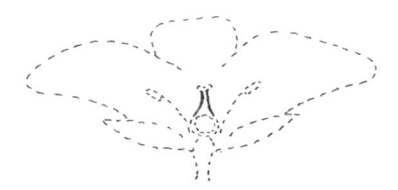

sub- - 1) A prefix added to a descriptive word to show that what you are looking for is underneath, e.g., a subterminal flower is below the end of the branch.

2) Also used to mean 'almost,' e.g., subglobose shape is almost round, like a tangerine, *Citrus reticulata*.

sub-shrub - A plant with some woody parts, usually roots and main stem, which dies back to the woody parts each year, then sends up fresh green shoots in the spring, as in the fig, *Ficus*.

subtend (sub tend') - To occur immediately below, as a bract or stipule subtends a flower stalk (peduncle) or leaf stalk (petiole).

succulent (suck' you lent)- Fat and juicy, like a cactus, *Mammalia,* or the leaves of marsh marigold, *Caltha palustris*.

sucker - 1) Shoot coming up from the roots, especially if roots are nicked by the mower. Also said of sprouts from the stump of a recently cut tree.
2) National Botanical Research states there is a new one born every minute.

superior - Located above something, like a superior ovary is above the petal attachment. In botany, above means farthest from the point of attachment, so if the flower is hanging down, the ovary inside the flower petals would still be above them, and therefore superior. Remember that much of the description of plants is done in the herbarium with dry specimens. Compare with inferior.

survey – 1) **Land survey**: Taken to prove boundaries by measuring angles, distances and elevations.
2) **Vegetative survey**: Precise methods determine the species and prevalence of each in a specified area. Synonym: vegetative analysis.

suture (sue' tyoor) - A seam or line, often referring to the line along which a fruit opens (dehisces).

swamp – A habitat that is underwater during the spring of the year, but dries out in the summer. See marsh, wetland.

symbiosis (sim' bee ose' iss) – The relationship between two entities where each gains something from the other. For example, a woodpecker and an old maple tree, *Acer*. The woodpecker finds food and a nest cavity in the tree, and the tree has the bugs removed from its broken limbs.

symmetrical (sim met' tree kal)- Having opposite sides that are identical. Imagine holding a string in front of your face—crossing the nose from chin to forehead—while you look in the mirror; on each side of the string, you see one ear, one eye and half of all the rest. This is **monosymmetrical** or more often called 'zygomorphic' (paired shapes), and refers to flowers like pansies, *Viola*. **Radiosymmetrical**, more often called 'actinomorphic' (shaped like light rays), refers to the face of a flower that is round with all the petals of the same size, shape and position, like a poppy, *Papaver*.

sympatric (sim pat' rick) – When two or more similar species inhabit the same range but do not interbreed.

synonym (sin' oh nim) – In botany, a synonym is a name that was once used as a binomial, but has been replaced and a different name approved by the Powers That Be (International Code of Botanical Whosis) as the 'correct' name.

The cloning of humans is on most
of the lists of things
to worry about
from Science,
along with
behavior control, genetic engineering, transplanted heads, computer
poetry and the unrestrained growth of plastic flowers.
—Lewis Thomas

T

tap root - A storage organ having a thick central root with thread-like roots growing off the sides, like a beet, *Beta vulgaris*.

taxon, plural **taxa** (tacks' on, tacks' ah) - Any unit of classification, like kingdom, division, order, family, genus, species, subspecies, variety, form, cultivar; or the largest group for beginners, unknown.

taxonomy (tacks on' oh me) - The science of identifying, describing, classifying and naming plants and animals.

tender – In reference to a plant that shivers in cold weather, and will probably die with even light frost. Compare with hardy, half hardy.

tendrils (ten' drills) - Modified leaves that have only the midrib left, and are adapted to twining around other plants and upstanding objects to climb to the sunshine.

tenuous (ten' you us) – Slender, thin.

tepal (to rhyme with people)- A part of a flower that looks like a petal because it is bright colored, but functions like a sepal in that it is the outer layer of the bud. The three outer 'petals' of a tulip, *Tulipa*, are tepals.

terete (tair eat') - 1) Referring to a cylinder such as a stem, either solid or hollow. 2) Round in cross section.

terminal (term'in al) - Referring to anything at the end (apex) of a branch or other central line (axis). Synonym: apical.

terminal bud - 1) The bud that forms on the very tip of a branch, with no leaf scar beside it. Synonym: apical bud.
2) What you begin calling a previous friend who won't help you with botany.
3) The last one in the six-pack.

terminal flower – Appears at the tip of a branch rather than between the leaves or branches. Compare with axillary flower.

ternate (turn' ate) – Appears in groups of three.

terrestrial (tair ess' tree al) – In reference to a plant that grows on land, not in water or in tree tops.

testa (test' ah) - 1) The outer coat of a seed.
2) The New England grad student who takes over for the professor when an exam is given.

tetra- (tet' rad)- A combining form meaning four.

Thallophyte (thal' low fight) - A flowerless plant from the great group Thallophyta, that includes algae, fungi and lichens.

thistle (thiss'ull) – A group of armed plants of Asteraceae family. See prickle.

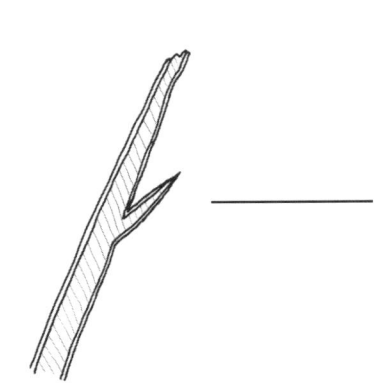

thorn - A sharp structure that grows from the woody layer, as in pear trees, *Pyrus*. Rose "thorns", *Rosa sp*, are actually prickles. See spine, prickle. See appendix - armor.

throat – In united flowers (corollas), the area between the attachment (tube) and the flared top (limb).

thugs – Invasive plants, weeds.

timber – Raw wood, fresh off the stump, or perhaps seasoned for a year after removal of bark. When cut and prepared for building, the name changes to lumber. See lumber.

tine – See prickle.

tissue culture – A relatively new process of vegetative reproduction. Parts of desirable plants are ground up into a soup (microcutting), then carefully placed on a sterile nutrient jelly to develop into cloned plants.

tissues - See organs.

tomentose (toe meant' toes) - Wooly, with soft matted hairs.

tooth, pl. teeth – 1) Serrated edges.
2) Occasionally used to describe armor of rigid prickles on stems or the margins of leaves like holly, *Ilex sp*.

top - The botanical top of a flower is the part farthest from the point of attachment. When a bellflower hangs down, the part closest to the ground is the top of the flower, because it is farthest from the stalk. Synonyms: apex, apical.

topsoil – The natural upper layer of soil, containing decayed plant material.

tortuous (tore' tyu us) - 1) All bent out of shape, like tendrils on a vine.
2) Having a surface that is twisted.

trailing - Referring to a stem that sprawls along the ground. Synonym: recumbent.

translucent (trans lou' sent) - Sort of see-through, like frosted glass. Lets light through, but not enough to see shapes clearly.

transpiration (trans' purr ay' shun) – The release of water vapor from the pores (stomata) of leaves. This is believed to increase the upward flow of water from roots.

transverse (trans verse') – At a right angle to the long stem (axis). Synonym: horizontal.

tree – A large perennial woody (ligneous) plant with a single stem (trunk).

tri- (try) – Combining form meaning three.

trifoliate (try foal' ee at) - Having three leaflets on a compound leaf. If the leaflet stalks (petiolules) come from the same point, it is called **palmately trifoliate**.

If the center leaflet stalk is longer than the other two, it is **pinnately trifoliate**.

tropism (trow' p ism) – The act of movement of any plant part in response to some stimulation, such as sunshine.

true - Seeds of a plant species always produce plants that look like the parent plant, except in the extremely rare case of a mutation. These seeds are said to come true. Seeds of hybrid plants may produce the crossbreed that you observe, or it may produce a plant that looks like the male (staminate) parent or the female (pistillate) parent which may be very different. We say the hybrid phenotypes do not come true.

true leaves – When a seed germinates, the first leaves to appear are the seed leaves (cotyledons). The next leaves are rudimentary leaves that will mature to the way the species should look. These are true leaves.

truncate (trunk' ate) - Ending abruptly, appearing to be cut straight across.

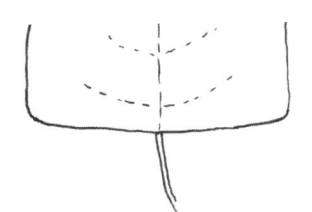

tube, adj. **tubular** (tube' you lar) - A hollow cylinder that is about the same size all the way up, as on a flower with petals grown together (united petal), though they may flare abruptly at the tip (limb).

tuber (two' bur) - 1) A thickened part of an underground stem that stores food and may start a new plant, such as white potato, *Solanum tuberosum*. A few tiny tubers may grow on above ground stems, and some books say tubers may grow on roots or stolons.
2) The instrument tooted by a strong member of the marching band.

tubercle (tube' er kull) – A tiny swollen structure, usually different in color or texture from the organ where it is found, a lump or nodule.

tumescent (tomb ess' sent) – Swollen, especially from internal pooling of liquid.

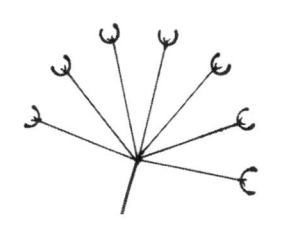

twining - 1) Twisted around something, like tendrils of grape vine, *Vitis*. See tendril.
2) Referring to a stem that coils around objects without help of tendrils, suckers, etc., like the stems of morning glory, *Convolvulus*.

type - When a new species is named, the original dried specimen is housed in an herbarium. All other plants that are judged to be of the same species are compared to that original dried "type" specimen.

U

ultraviolet [UV] (ull' trah vie' oh let [you vee]) – Part of the light spectrum that people can't see, but calls in night insects who can easily see in it.

umbel (um' bell)- A flower cluster shaped more or less like an umbrella. Many flowers on stalks (pedicels) coming from the same point on the main stalk (peduncle) shape into a flat or rounded top in a **simple umbel**.

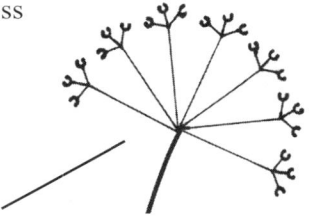

In a **compound umbel**, flower stalks called **rays** all come from the same point at the top of the main flower stalk (peduncle), but each tiny flower is on a smaller stalk called a **pedicel**.
Indeterminate umbels start blooming at the outer edge and the stalk continues to grow in the middle. **Determinate umbels** start blooming in the middle of the cluster and the stalk stops growing. See determinate, indeterminate.

undergrowth – Plants that grow in the shade of large mature trees. Also called **understory** and **underplantings**.

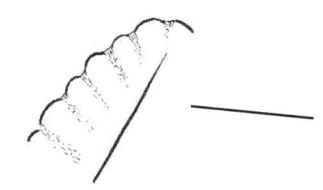

undulate (und' you lat)- Describing a leaf where the length of the margin is longer than the inner part of the leaf, causing the margin to be wavy.

uni- (you' knee)- A combining form meaning one, as unifoliate means one leaf.

unisexual (you' knee sects' you al) – Having only one sex in a flower; with no male parts (stamens), or alternately, no female parts (carpel, pistil), but still fertile. Synonym: imperfect. Antonyms: perfect, bisexual. See monoecious, dioecious.

united petal - A flower in which the petals are grown together at the base. If you try to pull one petal off, either the whole corolla will come loose, or it will tear.

(back view)

uptake – A plant's ability to absorb water from soil to utilize the nutrients the water contains.

urban (er' bun)- In reference to human habitat as contrasted to wild lands.

urn - 1) A round bowl with a flare at the top. From a Latin word meaning where urine is stored, so it looks like a chamber pot. A corolla of this shape is called **urceolate**. (Remember that the bottom of a drooping flower is technically the top.) See top.

2) What you hope to do financially when you finally get your botany degree.

utility - Uses for a plant, like windbreak, wildlife shelter, food or medicine.

V

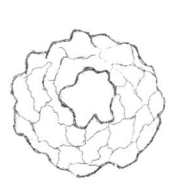

valvate (val' vate) – A pattern of joined petals (united) or sepals joined at the base (connate) all crumpled up in the bud with two bud scales, as shown in this cross section. Think of it as being like gathering a sock around your thumbs before you slide it over your toes.

variegated (var' ee egg gated)- Having two or more colors in irregular patterns.

variety - 1) All of the possibilities of differences within a species, such as populations where that group may have a different leaf shape or petal color, variability of certain traits such as the hairiness of the stems, character variation like a difference in seed shape, and many other conditions.

2) A commercial horticultural term for a plant with some difference, like a different streak of color on the blossom, and often legally patented. There are over 40,000 registered daylilies, *Hemerocallis*. Synonym: cultivar. See PPAF.

vascular (vas'cue lar)- Referring to the system of tubes (xylem and phloem) that carry water and food up, down, and around the plant. Similar to the blood circulatory system in people.

vegetative growth – Producing a new plant by dividing cells, without sexual reproduction. This could include cuttings, slips, runners, stolons, bulbs, tissue culture, etc.

vein (vane) – A vessel that carries fluids.

vein scar – See leaf scar.

venation (veh nay' shun)- The pattern of veins on a leaf: straight (parallel) in the monocots, and netted (reticulate) in the dicots, with a few exceptions.

ventral (ven' trull)- Attached to the front or inner surface. Antonym: dorsal. Illustrated at dorsal.

vernal (ver' nal) - Appearing in spring, especially blooming then.

vernation (vern nay' shun) – The placing of leaves in a bud.

verticillate (ver tiss' ih late, or lat)- Arranged in whorls, like leaves on a stem of tiger lily, *Lilium*. Synonym: whorled.

vesicular (vess sick' you lar) - Inflated with liquid or air.

vestigial (vess tij' ee al)- Said of a part that was complete in plant ancestry, but now has shrunk until it is no longer functional and often barely visible. Synonyms: obscure, rudimentary.

villous (vill' us)- Covered with long, soft, curly hairs that are not matted. Tiny hairs are referred to as **villosulous**.

vine - Any plant that climbs up other plants until it reaches sunshine. Vines climb by the use of modified leaves such as tendrils and suction cups, by adventive roots which penetrate the support plant, or by the true stem or leaf stalks twining around the host plant. Most vines prefer to germinate in shade, but grow upwards to have their crowns in sunlight before they blossom. There is much confusion in the literature regarding descriptions of various ways of climbing, using non-technical terms such like 'scandent' and 'clambering' as though they were technical terms. Asa Gray in 1858 *School and Field Book of Botany* listed 'scandent' as a synonym for 'climbing, using tendrils, suction cups or rootlets.' Some horticulture books still list 'scandent' that way, and list 'clambering' as climbing over things without special attachments, like low mallow, *Malva rotundifolia*, does. Many modern plant systematics books ignore the distinctions, and say they are synonyms, or do not list them at all.

viscid (vis' sid) - 1) Sticky, slimy, with a ropy consistency. Synonym: Glutinous. 2) Overlaid with a sticky layer, especially in leaves.

volunteer – A plant other than a weed that sprouts unexpectedly. The seeds may have blown in on the wind, be carried in a bird's intestines or tangled in your dog's fur.

W

water stress – Drought.

water table – The top of underground water in saturated soil.

weed – A plant for which no use has as yet been found. See pioneer plant, thug.

wet feet – Said of plants like cattail, *Typha*, that enjoy having their roots under water.

white rot – A fungal infestation of decaying woody materials, common in nature, and found to work slowly but efficiently in destruction of chemical pollutants.

whorl - A circle of leaves or branches around a stem, all growing from the same height at a node.

wildcrafting – A term used by herbalists to show they have been trained in how to collect wild herbs without harming the species.

wildflower – Any plant that grows and flowers in nature with no help, often also providing food and shelter to wildlife. See native plant.

wild type – The appearance (phenotype) that looks like plants of a species growing under natural conditions before selection, hybridizing, etc.

wilt – 1) Loss of firmness, drooping of plant parts due to insufficient water inside the plant. May be caused by disease, such as root rot from too much water.
2) A disease that causes collapse of leaves, usually an infection by a fungus or bacterium.
3) What you do when embarassed.

windbreak – A row of trees planted on the upwind side of a property to control air movement.

wings - 1) The back-sweeping petals (standard) on a flower like a pea, *Lathyrus*.
2) Flattened out-growths along the sides of a stem, seed or other part.
3) Fast food, *Bisona alatus*, often found in botanists' lunch bags.

winter annual – A plant whose seeds germinate in the autumn, survive winter, bloom and set seed in the spring, then die.

winter kill – Branches needing to be pruned that died from exposure to seasonal cold. See die back.

witch's broom – Dense clustering of branches of woody plants on an occasional branch because of injury or other factor.

wolf tree – An old tree with widely spread branches growing in an area of narrow trunked upright trees that sprouted after the wolf tree was already mature. The term 'wolf' refers to the fact that it uses nutrients and space that could be earning money for the forest. Destruction of the wolf tree occurs in most cases. See seed tree.

wood – The hard (lignate) secondary (not apical) tissue of trees and shrubs.

woody – Ligneous, almost as hard as wood, as the old stems of some perennial plants.

woolly – With many hairs covering the surface so it appears almost like felt. See lanate and tomentose.

wort – Any plant or herb in antique language, but especially a medicinal or edible herb.

X

X (ecks) - 1) A botanical symbol indicating that the parents of this plant were of different species that crossed, a hybrid. Common in oaks, *Quercus*, and violets, *Viola*. among many others.
2) The mark you make on a map to go back to collect mature seeds, only to find that wildlife had the seeds for lunch.

xanthic (zan' thick) – Yellowish.

xeric (zere' ick)- Growing in a hot, dry habitat. Compare with mesic.

xerophyte – A plant adapted to dry conditions, like a cactus, Cactaceae.

xylem (zye' lem) - 1) The water-bearing tubes of woody plants.
2) Wood necessary for the production of xylophones.

Y

yes - What you should say after an hour of lecture, when a botanist asks if you understand something, in order to avoid another hour of lecture which won't change things a bit in your overloaded brain.

Z

zone – 1) An area of different color on a flower.
2) A plan to differentiate geographical growing conditions, especially cold and frost limits. Zone 1 lays near the Arctic Circle with few species adapted, and Zone 11 is tropical. Good seed catalogs usually print a map of zones, and list the range of zones where each species/cultivar will thrive. Other nations have produced similar maps. Look for Australia (Australian Dept of Environment and Heritage), Canada (Agriculture Canada), and European Hardiness Zones. Other countries are following. Sunset Magazine developed a separate 24 zone map for the US.

zoophilous (zoe ah' fill us) – Pollinated by animals.

zygomorphic (zye' go morf' ick) - Refers to a flower that has a mirror image in only one direction, such as an orchid, Orchidaceae.

zzzzzzz - 1) Sound produced by a botany student during lecture.
2) Sound produced by various stinging insects who move faster than you can even when you are not making zzzz's.

**Study nature.
Love nature.
Stay close to nature.
It will never fail you.
—Frank Lloyd Wright**

*The sun,
with all those planets revolving around it and dependent upon it,
can still ripen
a bunch of grapes
as if it had nothing else in the universe
to do.*

— Galileo

		Evidence of its existence.
Filament and one-celled algae	1.6 billion years ago	
Lower Devonian period	395 million years ago	Scottish Rhynie chert deposits show *Rhynia*, one of earliest vascular plants in fossil record. Land plants became significant, where algae had been before.
Cretaceous period	136 million years ago	First evidence of flowering plants.
Fossil water lily (Nymphaeleale)	125-115 million years ago	Fossilized flower from Early Cretaceous of Portugal.
Two kinds of water lilies	66 million years ago	Fossilized leaves and pollen grains found in same time period as dinosaurs.
Lucy, pre-human remains discovered	Prehistory 3 million years ago	Hunter-gatherers use acquisition tools, such as digging sticks and crude containers like sea shells, skulls, tree bark, turtle shells.
Shanidar Cave, Iraq	50,000 years ago	Archeologists found date seeds, chestnuts, walnuts, pine nuts and acorns.
Man from Ferrassie	40,000 years ago	Improved acquisition tools, like bone scapula trowels and pottery containers to transport, preserve, cook and eat food. Used fire, and cooking brought community together.
North America	34,000 years ago	Homo sapiens migrations during recessions of the Ice Age.
Nile Valley, Egypt	17,000 years ago	Charred remain of 25 different plants, including nut sedge tubers, cattail roots and dates.
Global	10,000 years ago	Glaciers recede, seasons change, solar warming strong.
Modern man Homo sapiens sapiens	10,000 years onward	When they recognize that seeds grow plants, start gardens for food and medicinal plants, develop beginnings of horticulture, agriculture. Select foundation foods like wheat, oats, peas, rice, lentils. Develop food production tools, manual and animal-powered. Villages and towns grow in size, work divisions appear.
Northern Europe	8,000 BCE	Immigrants bring bow & arrow, fish traps, hooks into reindeer herder area. Flint sickles and grinding stones used.
Israel	6500 BCE	Evidence found of agriculture: Faba beans, lentil, pea, chickpea.
Peru	6000 BCE	Lima beans, green beans, chili peppers grown
Central America	5000 BCE	Gourds, squash, beans, chili peppers, and modern corn (maize).
China	5000 BCE	Domestic rice, cabbage seeds in earthen jars.
Native people, North America	5000 BCE	Live along rivers and cultivate crops. Domestication of some wild plants along Mississippi River.
China	4700 BCE	Mulberry trees grown to feed silkworms.
Britain	4500 BCE	Evidence of managed woodlands.
Pakistan	4000 BCE	Cotton seed; also grape seed from Afghanistan.
Thailand	3900 BCE	Rice planted and harvested.

Tigris, Euphrates, Phoenicia	3500 BCE	Boats, oars, sails, nets, commerce between areas.
World population	3000 BCE	Total population 100,000,000.
Sub-Saharan Africa	3000 BCE	Sorghum planted.
Iraq and China	3000 BCE	Written herbal remedies.
Fah Shen-Chih Shu, China	2800 BCE	Five sacred crops: soybeans, rice, wheat, barley and millet.
Saqqara Pyramid, Egypt	2750 BCE	Archaeologists excavated a coffin from this period, made of six layers of wood like plywood: cypress, juniper and cedar of Lebanon.
Sumerians	2500 BCE	Used sulfur compounds to control insects and mites.
Sub-Saharan Africa	2000 BCE	Pearl millet; and near Mediterranean, olives.
Native Americans	2000 BCE	Gardens with corn, squash, beans and sunflowers, as well as food and medicinal use of many wild plants.
Hammurabi	Ca. 1800 BCE	His Code speaks to the practice of hand-pollinating date palms, sections on irrigation canals and property laws for gardens.
Egypt	1550 BCE	65 foot long scroll lists about 800 herbal medicines, including anise, caraway, cassia, coriander, fennel, cardamom, onions, garlic, thyme, mustard, sesame, fenugreek, saffron and poppy seed. With no refrigeration, spices were used to preserve food.
China	Ca. 1500 BCE	Ancient pleasure gardens, with introduction of peony and rose.
Ramses II, Egypt	1300 BCE	Maintained apple orchards along the Nile.
King Tut	Ca. 1307-1325 BCE	The Boy King of Egypt, crowned at age 9, died at 18. An olive branch from his tomb is considered the oldest herbarium specimen.
China	1200 BCE	Botanical insecticides for seed treatment; also used mercury and arsenic for body lice.
Nebuchadnezzar II	Ca. 604-562 BCE	Legend says he built the Hanging Gardens of Babylon for his mountain-born wife as a love offering.
Anaximander Greece	Ca 550-600 BCE	Saw spontaneous generation as historical event, followed by transmutation and evolution. Maps of known world, sailing guides, travel times.
Hippocrates-Greece	440 BCE	Founded school of rational medicine, herbal uses. Wrote on many of today's spices as medicines.
Theophrastus	Ca. 300 BCE	Wrote *Historia Plantarum* and *De Causis Plantarum*, studied characters, morphology, external conditions, modes of generation. Inherited Aristotle's botanic garden, many of his treatises. Considered Father of Botany.
Maya	250 BCE	Evidence of intensive cacao plantations.

Carthage	203 BCE	Tribute to Rome included 500,000 bushels of wheat and 300,000 bushels of barley.
Virgil	Ca. 50 BCE	Agricultural advice included laying fields fallow, allowing legumes to mature in a field before planting wheat, and scattering manure and ashes on fields.
Marcus Pollio Roman architect	13 BCE	First recorded rat-proof granary.
World	7 BCE	Total population 250,000,000
Bible and other writings	Ca. 32 AD	Palm Sunday story mentions date palms. It was grown for sugar content, also believed to promote human fertility.
Pedanius Dioscorides Apparently an army doctor, then attending physician.	Ca. 50-60 AD	Compiled *Materia Medica* with 600 species, one of most widely used texts during Middle Ages. He preserved copies of Theophrastus and other early herbalists, originals now lost. Avoided superstition and hearsay. First recorded anesthetic. Called Father of Medical Botany
Pliny	Ca. 70 AD	His book, *Natural History*, covered 1000 different plants, was a major source of information on botany. He listed exact costs, and gave tips on practices like alternating years growing beans and spelt.
Pompeii	79 AD	Buried by the volcano Mt Vesuvius erupting. Uneaten walnuts were found on a table used by priests.
Alps	400 AD	Glaciers advance, Europe cools.
Holland and Germany	746 AD	Added hops to beer. British did not use hops until 1524.
Hindu and Arabic scholars	760 AD	Use decimal arithmetic, detailed costs of plants and materials, percentage of cuttings that grew, quarts of berries picked, market prices of produce.
Flanders and Zeeland	900 AD	Built dikes to hold back the sea from lowland areas to open land for farming. Holland did this 300 years later.
England	1046	Coldest winter in memory.
Saint Hildegard von Bingen	1098-1179	Greco-Roman men learned medicine by reading, women by doing. She was the first known woman to discuss herbal properties with men. Raised as a nun, she was consulted by and advised popes and kings. She wrote two major books, *Holistic Healing* and *Physica*, an encyclopedic work of materia medica in German, the first list of German names of plants.
Spain	1150	Made first paper in Europe, process brought by Moors.

Name	Date	Description
Albert of Bollstadt (Albertus Magnus)	1193-1280	Wrote *On Plants*, a temperate tone on medical values. Taught St Thomas Aquinas. Worked on morphology, said domestic plants can run wild, wild ones be domesticated.
China	Ca. 1200	Grew opium poppies.
St Francis of Assissi	1227	Mentioned as a holy man who loved animals and nature.
Europe	1250	Books mention the use of wheelbarrows for farms and gardens.
Thomas of Sarepba	Early 1300s	First known dried collection of plants, book form, glued to blank pages.
Medieval Europe	1315-17	Worst famine ever. Bad crop in 1315, people ate seeds for next year. 50% of livestock died, poor people starved.
William of Occam (Ockham)	1347	Occam's Razor: "What can be explained by the assumption of fewer things is vainly explained by the assumption of more things." Simplest answer is usually right. This is important in botany since scientists search for optimal solutions.
European plague	1350	Country gardens provide retreats for those escaping the disease that reduced the population by 60% before it ended about 1357.
Johann Gutenberg	1450	Invented moveable type for printing presses, significantly changed the world of ideas.
Berne, Switzerland	1476	Cutworms were taken to court, pronounced guilty, excommunicated, and banished.
Christopher Columbus	1492	Sailed west to try to reach spice lands in East. Returned with corn (Zea mays), strawberries, pumpkins. He made other trips, carrying seeds in both directions. In 1502 he captured a Mayan trade canoe, with seeds of cacao.
Diaz, deGama, others	Ca. 1500	Beans, sweet potato, and lima beans, natives to America, were brought to and used in Europe.
New World	1516	Bananas introduced from Africa.
Magellen	1519	Started from Spain with 5 ships. Only one returned in 1522, Magellen had died in 1521. Even with losses, 26 tons of cloves, plus nutmeg, mace, cinnamon and sandalwood made a profit.
New Spain (Central America)	1529	Report on Aztec gardens.
Otto Brunfels-Germany	Ca. 1530, Bern, Strassburg	Printed *Herbarum Vivae Eicones* (Living Images of Plants), described and illustrated 'large number' of plants of Central Europe. First link between mystic ancients and modern medica, quoted ancients, added observations.

University at Padua	1533	First professorship in botany, separate from department of medicine, established plant study as important in itself.
Hieronymus Bock-Germany (Jerome Boch)	1539	Printed *KreutterBuch*, in German for laymen to locate and identify medicinal herbs, perhaps the first 'home doctor book'. Arranged plants by resemblance, first attempt at natural classification.
Jacques Cartier	1541	Introduced cabbage to Canada, first US record was 1669.
Venice	1541	Cookbook for sugar published.
Leonard Fuchs-Germany	1542, Basel	Printed *Historia Stirpium* (History of Plants), with excellent woodcuts for identification. Indexed in German, Latin, Greek and common names. First mention of corn (maize, *Zea Mays*)
Luca Ghini, Botany professor, University at Pisa	1543	Europe's first botanical garden, medical plants.
Mattioli	1544	Wrote the famous herbal, *Commentarii*, which had at least 45 editions in many languages. Based on the work of Dioscorides, added all the plants Mattioli knew.
Italy	1550	Introduced from Mexico, tomatoes became important part of diet.
William Turner "The Father of English Botany"	1551	Printed in English and published the first part of *New Herbal*. Two volumes added in 1562 and 1568. Scorned superstition. This included much medicinal detail as to the worthiness of different plants for treatments, Used woodcuts of Fuchs.
Dodoens	1554	Produced 'Cruydeboeck', 900 page folio with 1309 woodcuts. Added plants of Netherlands. Condemned Doctrine of Signatures.
Clusius, France (LaEcluse)	1557	Published *Histoire des Plantes*. French translation of Dodoens works. Added plants of Austria, Hungary and Spain. Interested not only in medical uses, but in the plants themselves.
Andreas Caesalpinus, Italy (Cesalpino)	1583	Early attempt at Florence to make a methodical arrangement of the 1520 named plants into 15 classes by studying the fruits, instead of medicinal uses. Wrote *De Plantis*.
Carolus Clusius, Leiden (de l'Ecluse)	1593	Leiden established *Hortus Academicus*, in 1587, first botanical garden dedicated to ornamental plants. Clusius brought it to 5,864 species by 1750. He distinguished plants by leaves. Opened to the public in 1740, "couples openly in love are on no account admitted".
Europe	1594-97	Devastating famine struck, caused by 4 bad harvest years.

France, England	1595	Severe winter: Sea, olive trees, Rhone River all freeze, crops fail.
Bakers, Montpelier, France	1595	Forced to use bushes to fire ovens because forests were gone. Eventual reliance on coal, then petroleum, both from plant life.
John Gerard, a barber-surgeon, student of English plants	1597	Had a garden at Holborn, wrote and illustrated *Herbal of General Historie of Plants*, developed a crude system of classification based on the general appearance of plants and their uses.
Spigelius	1603	Published instructions for making dried herbarium specimens. Collecting, sorting and studying pressed dried plants mounted on loose sheets of paper changed taxonomy and systematics.
Jamestown colonists, Virginia	1607-1609	"Abundance of fish." Planted cucumbers and carrots in their gardens.
Lake Beemster, Holland	1612	43 windmills drained 225 square mile Lake Beemster to create 17,000 acres of fertile land. From 1550 to 1650, 400,000 Dutch acres of lake bottoms became farmland.
Gervase Markham, London	1615	Wrote *The English Huswife* "Contayning … vertues which ought to be in a compleat woman; As her skill in Physicke, Cookery, Banqueting-stuffe…, Perfume, Wooll, Hemp, Flax, Dayries, Brewing, Baking, and all other things belonging in a household."
Plymouth Colony Pilgrims	1621	Thanksgiving feast to thank Massoit tribe for their help in gaining the first harvest. Indian gift of corn in 1620 meant the survival of half of the original 102 Pilgrims.
Native Americans, Virginia	1622	Killed a third of the Virginia population of European settlers in retaliation for raiding Indian cornfields. Settlers cut trees and grasses back from homes so they could see attackers. US traditionally keeps huge lawns though there is no longer a problem.
Galileo	1624	Sent a microscope to Cesi, with a letter, "But I beg you to notify me of the most interesting things you observe."
Boston Common (Massachusetts)	1634	Created in downtown Boston as a place for city dwellers to graze their horses, cows and other livestock.
John Gerard(e)	1636	Wrote most famous English herbal, *Herball*. First description of potatoes, used as cheap food for sailors after conquest of Peru by Spaniards in 1536.
Dutch	1636	Occupied Ceylon, forcing villagers to furnish quotas of cinnamon, following the policy of previous Portuguese.
Boston	1646	First recorded recipes for green salads. Farmers averaged nibbling 17 kinds of plants each day, perhaps city life made them hunger for it.

North and South Carolina	1647	Rice introduced to farmers. Today CA, AR, LA and TX are main states growing rice.
Jean Baptiste Van Helmont	1648	Published *Ortus Medicine*, on plant nutrition from water; coined the word 'gas' and described properties of carbon dioxide.
Nicholas Culpeper	1649	Published herbal, *The English Physitian or an Astrologo-Physical Discourse of the Vulgar Herbs of this Nation... Whereby a man may preserve his body in health or cure himself being sick for threepence charge with such things onely as grow in England, they being most fit for English Bodies.* Many absurdities, but also sold many editions.
Coffee	1650	Arrived in England from New World.
New England	1652	First pine trees felled for British ship masts. By 1775, easy sources of wood for masts no longer available on the East Coast.
John Hull, Boston	1652	Established a New England mint. Coins bore image of willow, acorn and (largest issue) Pine tree.
Oliver Cromwell, England	1658	Died of malaria, refused to take quinine because it was recommended by Jesuits.
Robert Boyle	1660-78	Gas experiments; studied effects of combustion and respiration on the atmosphere.
Dr. Robert Morison, first professor of botany at Magdalen College, Oxford	1669	Classified plants, followed the lead of Caesalpinus, but expanded the classes to 18, based on whether they were woody or herbaceous, and working with flowers and fruits.
Nehemiah Grew	1672	Published widely illustrated volume on his studies of plant anatomy.
Antonio van Leeuwenhoek	1674	Polished short focal length lenses for microscopes; described protozoa, bacteria, etc.
London coffeehouses	1674	No women allowed. *Womens Petition Against Coffee* stated: "We find of late a very sensible Decay of the true Old English Vigour...Never did Men wear greater Breeches, or carry less in them of any Mettle whatsoever." The problem was blamed on coffee.
Maryland	1680	Colonists near starvation, complain they had to eat oysters to survive.
Nehemiah Grew	1682	Suggested sexual nature of ovules and pollen. Reported a conversation with Thomas Millington: they agreed that flower pollen represents the male element.
John Ray, Great Britain	1682	Developed a system thought of as beginnings of the natural approach used today, biological species. He separated non-flowering plants from flowering ones, which he then broke into monocots and dicots. Publication of *Methodus Plantarum Novae*.

Who	Date	Event
John Ray	1686-1704	Published three volume *Historia Plantarum*.
Rivinus	1690	Developed a classification system based on flowers.
Northern Europe	1693	Famine struck. By 1694, 10% of people in northern France died.
Rudolph Jakob Camerarus (Camerer)	1694	Published *De Sexu Plantarum Epistola*, a conclusive demonstration of sexuality in plants.
Botanical Gardens	1600-1700	Started for the cultivation of medicinal plants by universities where medicine was taught, but soon expanded. The first was at Padua in 1545. Jean Gesner, Swiss physician/botanist said that by 1700, Europe had 1600 botanical gardens.
J.P. Tournefort, France	1694-1700	Studied vegetable taxonomy, described 8000 species and sorted them into 22 classes, based mostly on the form of the flower, sorting them as herbs and under-shrubs as opposed to trees and shrubs. He published a systematic arrangement.
Jethro Tull	1701	Invented the first agricultural machine, a seed drill.
Robert Hooke	1705	Posthumous publication of his *Discourse on Earthquakes*, speculation on the quakes moving fossils around.
American settlers	Mid 1700s	Dig deep ditches to drain swampland to increase their farmable land.
Carl von Linne (AKA Carolus Linneaus)-Sweden 'Father of Taxonomy'	1707-1778	The sexual system (surprise!), based on the pistils and stamens of the flowers. His most important contribution was the Binomial System, having only two words for each species, the genus and the species.
Thomas Fairchild	1715	Reportedly produced the first artificial hybrid plant.
Charles Townshend	1715	Popularized 4-field crop rotation: turnips, wheat, barley, clover.
Cotton Mather	1716	First certain account of plant hybridization in a letter discussing the 'infection' of Indian corn planted near yellow corn. Arguments over sexuality of plants brought many hybrids by 1760.
Stephen Hales	1727	Announced that plants gain nourishment from air, and studied how water rises in plants.
Farmers	1732	First record of crop seeds being planted in rows to make it easier to kill weeds.
John Kay	1733	Invented the fly shuttle, speeding cloth making, thus mechanizing weaving, though thread spinning was still a cottage industry.
Potato famine, Ireland	1739	Half a million Irish died from widespread potato crop failure.
Game of cricket	1744	Rules established in England. White willow, *Salix alba var caerulea* is preferred wood for cricket bats, from 15 year old trees, harvested at 65 feet (20 meters) tall.
James Lind-England	1752	Raised interest in using fresh fruit to prevent scurvy.

Carolus Linneaus	1758	Published *Systema Naturae*, setting out the concepts used by taxonomists today.
Sir James Edward Smith, England	1759-1828	Gained possession of the Linnean plant collection, and wrote *English Flora*, which was taught in Britain for many years, even after France and other European countries had abandoned it..
British crown land grants	1761	Reserved the best pine trees for masts only on British vessels. Angered colonists, first Revolutionary flag bore image of pine tree.
Kolreuter	1761	Observed insects role in pollination; gave detailed descriptions of relationships of insect activity and plant structure. This formed the basis for the work of Sprengel.
Linnaeus, under name CN Nelin	1763	Entered an essay in a contest (disagreement on whether he won) on freeing orchards from caterpillars with mechanical and biological controls. His birth name was Carl Nilsson, the Latin ending showed his education, CN Nelin was made of letters in real name.
Michael Adanson	1763	Pushed for taxonomy based on shared characters rather than evolutionary patterns.
James Hargreaves	1764	Invented spinning jenny to rapidly produce thread. Textile production rose from 2.5 million pounds in 1760 to 22 million pound in 1780s.
San Diego CA, mission	1769	Sweet oranges thrived. First large citrus orchard established at San Gabriel Mission in 1804.
Dutch traders, Amsterdam	1770	Destroyed an entire year's supply of nutmeg and cloves. Short supply raised prices enough to create fortunes.
James Cook, British Navy	1770	Scientific mission, started in 1768 to Australia, with naturalists Joseph Banks and Daniel Solander (pupil of Linneaus) for huge collection of new plants. Sting Ray Harbor was renamed Botany Bay.
Joseph Priestly	1770-74	Discovered "dephlogisticated air" needed for animals to survive and produced by plants. Antoine Lavoisier named it oxygen.
Joseph Banks, returned from Australia	1772	Named scientific advisor for the Royal Gardens by King George III.
Joseph Priestly and Jan Ingenhousz	1772	Investigated photosynthesis.
Jan Ingenhousz	1779	Published *Experiments on Vegetables*, showing that plants need light to produce oxygen, and they use carbon dioxide in the process.
Thomas Jefferson	1781	First American to grow tomatoes, also known as 'love apples'.
Jean Senebier	1788	Proved that it is sunlight, not the heat of the sun, that fuels photosynthesis.

Bligh, Captain of *Bounty*	1789	Thirsty sailors mutinied and threw overboard 1000 breadfruit plants, which required fresh water to survive the trip from Tahiti to the West Indies. Bligh was dismissed.
British Navy	1795	First sent limes on ships to prevent scurvy, though Lind had discovered scurvy cure in 1752. English sailors called 'limeys'.
Jan Ingenhousz	1796	Understood that plants carry on respiration with photsynthesis, using carbon dioxide from the air in nutrition.
Antoine de Jussieu, France	1789	Combined Linnaeus's Binomial System with Adanson's Natural System, to provide a workable classification for naming and grouping plants, the basis for today's studies.
French Revolution began	1789	Mass urbanization from the Industrial Revolution forced residents to move to the cities looking for work. The hungry and unemployed caused massive bread riots when prices doubled. Peasants suffered most. Lavoisier was executed in 1794.
Johann von Goethe	1790	Published how he thought leaves and floral parts had similar origins. His book became the basis of today's morphological theory.
Herbaria	1700-1800	Specimens were mounted on loose sheets rather than books, to be sorted in different ways as ideas about classification change.
Christian Sprengel	1793	His original drawings on different modes of pollination were ignored until Darwin confirmed his work in 1876.
Eli Whitney	1793	Invented cotton gin. Production rose from 3000 bales to 4.5 million bales by 1860. Increased slavery; depleted soils.
French government	1795	Offered ƒ12,000 to devise a method of preserving food for military use. Canning in glass bottles won by Nicholas Appert in 1809.
Robert Thomas Malthus	1798	Malthusian theory: "Population increases in geometric ratio, means of subsistence increase in arithmetic ratio." Too many people, not enough food.
US agriculturists	1799	Described sweet corn, traditional food of Iroquois tribe. Not easily accepted, but by 1980 it rated #1 US canned vegetable.

They say that time changes things, but you actually have to change them yourself.   –Andy Warhol

The future is something that everyone reaches at the rate of sixty minutes an hour, whatever he does, whoever he is.
--C. S. Lewis

I've been on a calendar, but I've never been on time.
--Marilyn Monroe

The times they are a-changing.
--Bob Dylan

First Industrial Revolution	1800-1870	Coal mining, railroads, and clearing land speed up greenhouse emissions. Better agriculture/sanitation speed population growth.
Karl Friedrich Burdach	1800	Coined the term 'biology' to the study of human morphology, physiology and psychology.
Harvard	1801	Opened its first botanic garden.
John Chapman	1801	Known as Johnny Appleseed, begins planting apple trees in Ohio Valley, and across the US.
John Wedgewood Uncle of Charles Darwin	1802	Worked with Joseph Banks and William Forsyth, George III's gardener, to lay foundations for the Royal Horticultural Society, started 2 years later.
Gottfried Treviranus and Jean Baptiste de Lamarck	1802	Separately broadened the meaning of biology to encompass all living things.
Charles Francois de Mirbel	1802	Concluded from observation that "Plants are made up of cells, all parts of which are in continuity and form one and the same membranous tissue."
Shakers, Enfield CT	1802	First recorded sale of seeds in small paper envelopes.
Nicholas-Theodore de Saussure	1802	Published experiments with the first methods for measuring photosynthesis; developed first balanced equation for the process.
Royal Horticultural Society	1804	Established in England.
American and European traders	1804	Stripped Pacific Islands of sandalwood to take to Europe and China. All sandalwood gone from Fiji in 1809, Marquesas by 1814 and Hawaii by 1825.
Ludolf Christian Treviranus	1805	Recognized that sperm were equal to plant pollen.
Lewis and Clark arrive at Columbia River	1805	New era of western expansion arrives. Lewis spent 9 months studying botany to prepare for the trip. Many plants collected by the group, seeds taken back.
Napolean	1806	Offered 100,000 francs to anyone who could create sugar from a native plant.
Jean Baptiste de Lamarck	1809	Studied plant (and animal) structure under the microscope. He recognized that cells are the basis of all life "and without this tissue no living body would… exist, nor could it have been formed."
Nicholas Francois Appert French chef/bacteriologist	1809	Invented the procedure of food preservation by canning.
Robert Brown	1810	Began publications on flora of Australia.
Robert Brown	1815	Classified higher plants, sorting angiosperms and gymnosperms.
Food riots	1816	Widespread crop failure in Europe, rioting in Eng., Fr., Belgium.

James Hart Stark	1816	Took apple tree scions from KY to MO, the foundation of the Stark Nursery, which remained in the family for over a century.
Agriculture experts	1820	Myths about tomatoes (known as love apples) shot down.
John Goss	1822	Reported segregation of a recessive trait in peas, but didn't catch numerical ratios.
Charles MacIntosh	1823	Waterproofed fabric by treating with rubber to protect outdoor workers in rainy weather.
Thomas Andrew Knight	1823	Proved dominance, recessivity and segregation in peas, but did not find regularities.
Pierre Jean Francois Turpin	1826	Observed cell division in algae.
Adolphe Brongniart	1828	Published complete record of known fossil plants, called Founder of Paleobotany.
Edward Beard Butting English textile engineer	1830	Invents lawn mower, which was imported to US 25 years later.
John Deere	1830	Invents and produces the first cast iron plowshare.
Hugo von Mohl	1835-39	Described mitosis in plants. "Cell division is everywhere easily seen...in terminal buds and root tips."
John Deere, Illinois blacksmith	1837	Invented the highly polished steel plowshare to turn sticky prairie soils. Established Deere & Co in 1868.
Rene Joachim Henri Dutrochet	1837	Recognized that chlorophyll is needed for photosynthesis.
USDA	1839	Developed library of agriculture.
John Dresser, MA	1840	Made a hand powered lathe to produce thin sheets of veneer, which led to production of plywood.
Justus von Liebig	1840	Found organic compounds in plants are synthesized from inorganic carbon dioxide in air.
Friedrich Keller	1840	Patented a wood grinding machine that made papermaking from wood pulp possible. Experiments produced coffins, horseshoes and road surfaces from wood pulp.
George Perkins Marsh, Congress, Vermont	1840s	Advocates a conservationist approach to forest management, warning of the destruction of human impact on the land.
Potato blight	1840	Hit Ireland, England and Belgium.
Ireland	1841	Potato famine begins, 1 million die, another million survivors emigrate to US.
Johann Japetus Steenstrup	1842	Described alternating sexual/asexual generations in plants/animals.
Rothamsted Experimental Station, England	1843	First agricultural experiment station in the world.

John Mercer	1844	Invented mercerization of cotton, stretching fibers under cold caustic soda; gives sheen, long life and better dye uptake.
Christian Schonbein, Germany	1846	Mixed sulfuric acid and saltpeter, splashes dissolved his cotton apron. When dried, it exploded: cellulose nitrate also called gun cotton, used in antique movie film and an ingredient of celluloid.
Asa Gray	1848	Published *The Manual of Botany of the Northern United States.*
US Congress	1849	Established Department of the Interior.
US Congress	1849	Passed Swamp Lands Act, allowing destruction of wetlands, granting Louisiana reclamation for all Federal lands.
Thames River England	1850	Last salmon taken from the river, salmon do not return to river for 150 years due to heavy pollution.
US agriculture Mechanization	1850	Opened a billion new acres by 1920, another billion by 1980. Shipping, refrigeration and processing all improved. Today's farmer gets 4% of price of chicken, 12% of a can of corn.
Joseph Hooker	1851	Brought 6,500 species of plant from India to Kew Gardens.
London	1851	Beginning 1820, lasting 80 years, "the baked 'tato man" sold hot baked potatoes from fall to spring. Over 300 vendors sold ten tons of potatoes each day.
Jean Baptiste Boussingault	1851-55	Demonstrated that higher plants cannot use nitrogen from air, only nitrates from soil.
Henry David Thoreau	1854	Publishes *Walden, or Life in the Woods.*
Technology	1855	Allowed production of rayon and cellophane from wood chips.
Mason jar	1858	Encouraged home canning; white sugar usage doubled between 1880 and 1915.
Wendlin Grimm, MN	1858	A handful of alfalfa seed, selected for resistance to winterkill, grew this over many years. In 1900 MN Ag Experiment Station grew it and released it as a variety.
Baseball bats	1859	The game started in the 1850s, with no regulations on bats. Now limited to 2.5" diameter, any length, even flat was OK. In 1869, max. length of 42" was set. White ash, *Fraxinus sp.*, became the favorite, then metal, but pros often prefer wood. In 2001, Barry Bonds broke annual home run record with 73. He was using maple bats, and sales of those soared.
James Caird, Briton	1859	Published *Prairie Farming in America*, said Chicago exported 100 bu of grain in 1837, over 2 million bu in 1847, 18 million in 1857.
Yearbook of Agriculture	1860	Stated a pint of clover from England contained 70,000 weed seeds.

Julius von Sachs	1862	Proved that starch was a product of photosynthesis.
US Congress	1862	Established a Commissioner of Internal Revenue, to pay for Civil War. Whiskey taxed at 20 cents a gallon, rose to $2 by 1865.
Abraham Lincoln	1862	Morrill Land-Grant Act authorized colleges to teach agriculture and mechanic arts, 13 million acres granted to states to support the Act.
US Congress	1862	Homestead Act passed, allowed people to farm vacant land for ownership.
Lincoln, US Congress	1863	Charters Union Pacific railroad, subsidized track miles. Human food and animal fodder can be transported.
US Congress	1864	Passes a bill granting Yosemite Valley as a public Park.
Jack Daniels	1866	Set up the first registered US distillery. His birthday is not certain, but he was between 16 and 20 years old when it started.
Gregor Mendel	1866	Proved that traits pass from parents to offspring, foundation of today's genetics. Results were lost for 34 years.
Earnst Heinrich Haeckel	1866	First used word "ecology" to describe study of interactions between living organisms and their environment.
Horticulturist Parker Earle	1866	Shipped strawberries from southern Illinois to Chicago by rail in iced boxes on Illinois Central Railroad.
Jean Baptiste Boussingault	1868	Proved plants require oxygen for photosynthesis.
William Davis Detroit Michigan	1868	Patented refrigerator car, actually ice-salt mixture, not mechanical.
USDA	1868	Created Division of Botany to preserve herbarium material from government expedition collections.
Railroads	1869	First transcontinental US railroad completed.
Iowa	1870s	Red Delicious apple discovered.
Carbon dioxide levels	1870	Early atmosphere readings find 290 parts per million (ppm), later confirmed in ancient ice.
Second Industrial Revolution	1870-1910	Chemical fertilizers, electricity, public health, transportation increase population growth.
Lambert Adolphe Quetelet	1871	Realized the importance of statistical analysis, and began the science of biometry.
US Congress	1872	Establishes Yellowstone National Park.
US Congress	1875	Bans unauthorized cutting of trees on government property, thereby stopping lumbering interests from harvesting free timber.
US Dept of Agriculture	1875	First ag experiment station established in CT. In 1876, they opened the first US seed testing lab.

Charles Tellier, France	1876	Builds ship, *LaFrigorifique* (The Refrigerator) and transports meat from Buenos Aires to France, introducing world transport of perishable foods.
Frederick Law Olmstead	1878	Begins work on Boston's "Emerald Necklace", a series of public parks surrounding the city. Boston Public Garden becomes the first botanical garden in the US.
Daniel Peter	1879	Using Henri Nestle's powdered milk formula, Peter made first milk chocolate bars.
US Congress	1879	Establishes US Geological Survey as a bureau of the Dept of the Interior.
Charles William Cooper	1880	First mechanically refrigerated railroad car patented in US. However, ice remained the standard for shipping.
Railroad and Seaway network	1880	Major continents connected by sea for world food transport (5 million tons total transport) and railroads cross the continents.
Bordeaux University	1882	Sprayed grapes with copper sulfate to discourage children from stealing them, noticed that the spray deterred downy mildew.
Eden Park, Cincinnati	1882	50,000 people attended first American Forest Congress.
Wilhelm Engelmann	1882	Discovered that red light most effectively promoted photosynthesis.
Emil Christian Hansen	1885	Introduced pure cultures for starting beer fermentation.
Woods Hole, Mass	1886	First Biological Station established.
Heinrich Waldeyer	1888	Found and named chromosomes.
Unnamed physician St Louis MO	1890	Concocted peanut butter as a high protein snack for invalids and those with poor teeth.
US Congress	1891	Passes the Forest Reserve Act, the legislative foundation for creating the National Forest system.
US Supreme Court	1893	Declared the tomato to be a vegetable, as opposed to the word 'fruit'. Tomato importers were required to pay a 10% vegetable tariff on imported tomatoes.
William Bateson	1894	Wrote *Materials for the Study of Variation* showing importance of discontinuous variations, bringing about rediscovery of Mendel's work.
USDA	1898	Plant exploration trip to Russia, brought back new durum and hard red wheat collections to grow in US. In 5 years, wheat production went from 60,000 to 20 million bushels.
Charles Reid Barnes	1898	Proposed the term "photosynthesis".
USDA	1898	Introduced avocados from Mexico.

US Pres Benjamin Harrison	1893	Sets aside 13 million acres of forest reserve.
International Congress of Genetics	1899	First world wide conference held in London.
Scientists	1900	Recognized only two kingdoms of living beings (plant and animal), with about 100,000 species of plants.
USDA	1900	Collected spinach in Manchuria that saved Virginia from spinach blight disaster.
Hugo DeVries (Holland) Karl Correns (Germany) Erich von Tschermak-Seysenegg (Austria)	1900	Worked separately and each claimed to have discovered and verified Gregor Mendel's principles, the beginning of modern genetics.
USDA	1900	Collected rice varieties in Japan, now major crops in Texas and Louisiana.
Motorized agricultural machines	1900 to present	One man is capable of feeding many. Farmers drop from 97% of US population to less than 2%. Malnutrition and starvation remain in undeveloped countries.
Wilhelm Johannsen	1903	Introduced and defined concepts of phenotype, genotype and selection.
Carl Neuberg	1903	Introduced the term "biochemistry".
Bronx Zoological Park	1904	Chestnut blight first detected, that eventually destroyed nearly all of American chestnut trees. Loss of lumber alone was estimated at $400 billion.
American Horticultural Society	1904	First established
US Congress	1905	Established Forest Service under Dept of Agriculture.
US Pres Theodore Roosevelt	1905-07	Sets aside 180 million acres for wildlife refuges and national parks.
Ag Experiment Station-Texas	1906	Collected cotton in Acala Valley, Mexico, grown today in Calif.
Arbor Day	1907	Based on Nebraska's "Tree Planting Day" from 1872, the US establishes and celebrates Arbor Day.
Western ranching/mining interests	1907	Organize opposition to US conservation policies. Call for end to restriction of free use of national forests.
American farmers	1910	One farmer could produce enough food to sustain 7 people.
Corn Products chemist	1910	Discovered process for refining corn oil for cooking, Mazola.
US Congress	1911	Passes Weeks Bill allowing acquisition of forest reserves to protect watersheds, mostly due to pressure from American Forests group.
Kudzu	1911	Brought to US from Japan for erosion control and livestock forage. Today it covers 7,000,000 acres of the South.

Frederick Hopkins	1912	Discovered substances - beyond fats, carbohydrates and minerals – essential to humans. Casimir Funk named them 'vitamines'.
Sea transport, world-wide	1913	12 million tons
Mosquito control	1915	Panama Canal was abandoned in late 1800s because of malaria and yellow fever. Control allowed completion.
Charles E Bessey	1915	Published *The Phylogenetic Taxonomy of Flowering Plants*. He trained under Asa Gray, became Professor of Botany at U of Neb.
US Congress	1916	Created National Park Service as division of US Dept of Interior.
Joseph Grinnell	1917	Defined the concept of ecological niche.
Ford Motor Company	1917	Introduced the Fordson tractor, at $397.
Thomas Hunt Morgan and coworkers	1919	Published *The Physical Basis of Heredity*, a summary of the rapidly expanding field of genetics.
Otto Warburg	1919	Found that photosynthesis was more efficient in intermittent light.
US Congress	1919	Prohibition of alcoholic beverages.
First aerial application of insecticide	1921	Against Catalpa sphinx moth in Ohio.
Otto Warburg	1922-23	Invented manometric apparatus to measure quantum efficiency in photosynthesis. It became the standard tool for measuring metabolism in living cells.
Thorsten Ludvig Thunberg	1923	Showed that carbon dioxide was reduced and water was oxidized when photosynthesis occurred.
Hans Molisch.	1925	Proved oxygen could be produced by shining light on preparations of dried leaves.
Mechanically refrigerated trucks	1925-30	Introduced for home delivery of milk, but soon added store delivery of fruits and vegetables as well as milk.
Albert von Szent-Gyorgyi	1928	Found hexuronic acid identical to Vitamin C and proposed the name ascorbic acid.
Sanforizer Co	1930	Introduced Sanforize process to prevent shrinkage of cotton fabrics.
Cornelius Van Niel	1930	Recognized parallel of photosynthetic process between bacteria and green plants.
US Congress	1930s	Encourages farmers to drain wetlands for farming and urban development by agreeing to pay half the costs. The Everglades of Florida loses large tracts.
M. Kroll and Earnst August Friedrich Ruska	1932	Built the first electron microscope

Texas Dept of Transportation	1932	Hired Jac Gubbels, landscape architect, to maintain, preserve and encourage wildflowers along rights of way. Highway patrol gathered wildflower seeds for germination. Today State Wildflower Trail is a major tourist attraction.
Franklin D Roosevelt	1933	Creates Civilian Conservation Corps (CCC) where young men from impoverished US families accomplished public works like planting forests, developing camps and tourist attractions.
All-American Selections	1933	First competition for best new flowers and vegetables.
US Congress	1934	Recognize value of wetlands, pass the Migratory Bird Hunting Stamp Act, collect fees to pay for habitat conservation projects.
John Deere	1934	Begins production of gasoline engine tractors.
US Congress	1935	Passed Rural Electric Admin Act (REA) and established Soil Conservation Service (SCS) under USDA.
A Koehler, botanist	1935	Analyzed wood from ladder used in the kidnap/murder of the son of Charles and Anne Morrow Lindbergh, same as planks in attic of Bruno Hauptmann's house. This was key evidence in the trial. Hauptman was electrocuted April 3, 1936.
A.G. Tansley	1935	Formulated concept of varied ecosystems.
George Washington Carver Tuskegee, Alabama	1936	Extensive work on peanuts, developing over 300 market products. He specialized in new uses for old crops. His work on 115 recipes for tomatoes is still used. His home became a National Park.
International Harvester	1938	Produced first Farmall gasoline tractors. IH tractors appeared in 1922 at $670.
DDT	1939	Recognition as insecticide.
Kausche and Earnst August Friedrich Ruska	1940	Published first electron microscope pictures of chloroplasts.
American Forests group	1940	Established National Register of Big Trees.
Gustaffson and coworkers.	1941	Produced new strains of cereal grains by selection from x-ray mutants.
United States	1941	Enters WWII after bombing of Pearl Harbor. Imports from overseas like coffee, tea, pepper, pineapple and orchids severely limited.
American population	1942	Produced 40% of all US vegetables in home Victory Gardens.
Plant breeding program	1942	Bred wheat successfully for resistance to Hessian fly.
Chemical companies	1943	Introduction of powerful insecticides: DDT, parathion, without understanding the adverse effects on the environment.

Entity	Year	Description
Great Britain (RHS?)	1945	Developed a seed mixture of four species that would survive under adverse conditions to revegetate bombed areas. Camomile was one.
Sweden	1946	First reports of insect resistance to DDT in houseflies.
American population	Late 1940s	Veterans of WWII return home, baby boom, flee cities, flock to suburbs, many forests cut down or fragmented.
Thor Heyerdahl	1947	Sailed a balsa raft from Peru to Polynesia to prove natives could have. In 1969, he sailed a reed ship and in 1970 a papyrus one.
American farms	1950	Tractors outnumber horses.
Francis Crick and James Watson	1953	Discover structure of DNA, perhaps most important biological event of the 20th century.
Brown, Bonner, Weir	1954-57	Estimated that if humans would eat products of algae and yeast, the Earth could carry 50 billion people.
Dutch Elm Disease	1950s and 60s	US urban forests destroyed, streets no longer tree lined.
National Seed Storage Laboratory	1958	Main US seed bank established at Fort Collins CO, along with 19 others. Ft Collins stores over a quarter million seed samples.
Mechanical refrigerated rail cars	1958	"Reefers" utilizing diesel powered refrigeration units are marketed. The last ice-cooled units were retired in 1971.
USDA	1960	Collected wild oats in Israel that helped breeders develop best-known disease resistant oat variety.
Sex pheromone	1960	Experiments showed the hormone attracted gypsy moths so they could be destroyed.
David Keeling Carbon dioxide levels	1960	Annual measurement of CO2, finds consistent yearly rise, now at 315 ppm.
Rachel Carson	1962	Published *Silent Spring*, exposed hazards of the pesticide DDT, made people face the inherent dangers in technology, set the stage for environmental movement. USDA ridiculed the book.
Integrated Pest Management (IPM)	1967	Term introduced, with relevance to ecology.
Earth Day	1970	First national celebration in US on April 22.
DDT	1970s	Widespread banning of the insecticide.
Mariner 9 spacecraft	1971	Finds dust storms warming Mars; greenhouse effect pushes temperature of Venus above water's boiling point.
US Environmental Protection Agency (EPA)	1972	Passes Clean Water Act, Section 404 requiring a permit to discharge dredge and fill material into the nation's waters.
Environmental Protection Agency	1972	Banned use of DDT in the US. However, it was still produced and sold to other countries.

Rafael Guzman, student U of Guadalajara	1978	Discovered a stand of ancient perennial corn (related to teosinte) that had survived in the mountains near Jalisco.
Liquid balsam	1979	Produced by copaiba trees so close to diesel fuel that diesel engines will run on it with no treatment.
US National Academy of Science	1979	Reports doubling CO2 will bring 1.5 to 4.5 C. global warming.
Reagan Administration US	1980s	Funding for urban forestry and other ecology programs cut. "If you've seen one redwood tree, you've seen 'em all." R. Reagan
Lady Bird Johnson and actress Helen Hayes	1982	Establish National Wildflower Research Center, Austin, TX, later Lady Bird Johnson Wildflower Ctr, united with U of TX in 2006. Mission: educate on value of native plants.
One American farmer	1982	Could produce enough food to sustain 78 people.
John Thomas, Texas	1983	Started Wildseed Farms, harvesting and marketing seeds from native plants, wildflowers that other farmers called weeds. Today they service Highway Depts in 29 states for roadside beauty.
Kary B. Mullis	1983	Devised a polymerase chain reaction, capable of increasing the sample size of DNA for forensics and paleobotany.
Iron sulphate	1986	First selective herbicide to kill broad-leaf plants.
American Forests group	1986	Famous and Historic Trees program started.
Germany, Indonesia, Phillipines	1986	Governments make IPM official requirement. Denmark and Sweden followed the next year.
Carbon dioxide levels	1988	350 ppm
United Nations	1988	Established Intergovernmental Panel on Climate Change (IPCC)
US Congress	1990	Passes Urban and Community Forestry Assistance Act, expanding funding twentyfold.
Resistance to chemicals	1993	At least 504 insect species show resistance to one or more formulas; 17 resistant to all classes; five kinds of rats resistant to controls; over 100 weed species resist herbicides.
Atlanta, GA	1996	First satellite tree cover analysis conducted.
Jean-Michel Cousteau	1999	Founded Ocean Futures Society, with Pacific island restorations a major thrust. Elimination of trash, rats and alien plants allows native plants and animals to thrive.
Malheur National Forest Oregon	2000	Discovery of world's largest organism, honey mushroom, 3.5 miles across, takes up equal of 1,665 football fields. Age between 2400 and 7200 years old.
Carbon dioxide levels	2005	380 ppm

Time flies like an arrow; fruit flies like a banana. –Groucho Marx

A committee is a group that keeps minutes and loses hours.
Milton Berle

Time is God's way of keeping everything from happening at once. –Albert Einstein

Look at the trees, look at the clouds, look at the birds, look at the stars... and if you have eyes, you will be able to see that the whole existence is joyful. Everything is simply happy. Trees are happy for no reason; they are not going to become prime ministers or presidents and they are not going to become rich and they will never have any bank balance. Look at the flowers – for no reason. It is simply unbelievable how happy flowers are. — Osho

<u>A tree growing out of the ground
is as wonderful as it ever was.
It does not need to adopt
new and startling methods.</u>
—Robert Henri

THE TROUBLE

WITH GARDENING

IS THAT IT DOES NOT REMAIN

AN AVOCATION,

IT BECOMES AN OBSESSION. —PHYLLIS MCGINLEY

*When I go into the garden with a spade, and dig a bed,I feel such an exhilaration
that I discover
that I have been defrauding myself all this time
in letting others do for me
what I should have done with
my own hands.*
—Ralph Waldo Emerson

It is difficult to place a monetary value on the many vital services that trees provide. However, the California Department of Forestry and Fire Protection calculates that a single tree that lives for fifty years will contribute service worth nearly $200,000 (in 1994 dollars) to the community during its lifetime. This includes providing oxygen ($31,250), recycling water and regulating humidity ($37,000), controlling air pollution ($62,500), producing protein ($2,500), providing shelter for wildlife ($31,250), and controlling land erosion and fertilizing the soil ($31,250). –Sacred Trees

<u>If I thought
I was going to die tomorrow,
I should nevertheless plant a tree today.</u>
—Stephan Girard

A GARDEN
IS NEVER SO GOOD
AS IT WILL BE
NEXT YEAR.
—THOMAS COOPER

PART III

TECHNICAL NAMES

WHAT BOTANISTS CALL
SPECIFIC EPITHETS
(SPECIES NAMES)

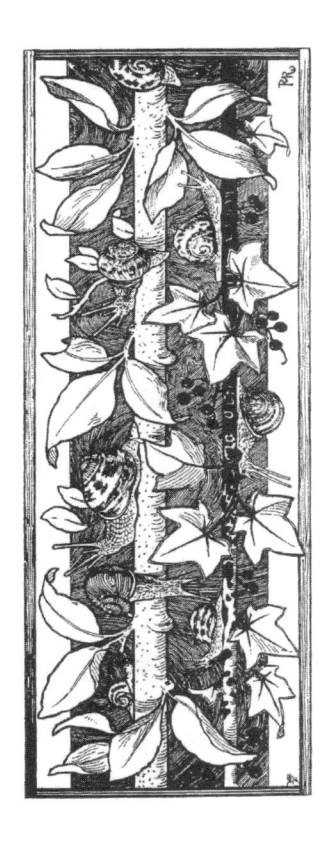

Words that are understood
all over the world
so everyone is talking about
the same thing.

L is for lanceolata

C is for cochleatus

R is for riparium

TECHNICAL NAMES:
Definitions.

When you located a picture of a wildflower you just found, the field guide probably listed a common name followed by two words that were all Greek to you. You ignored the last two words because you didn't know what they meant, and certainly had no idea how to pronounce them.

Those two words are called a "binomial", bi- meaning two, and nom- meaning name; a two word name that is recognized all over the world as referring to that plant. You can email a nursery in Sweden, or a friend in Japan and be understood about what plant you mean.

Binomials were invented to allow botanists, gardeners, herbalists, carpenters — and anyone else who works with plants or their products — to communicate through letters (there were no telephones then, to say nothing of fax and email) to be sure they were referring to the same plant.

For example, maybe a Swedish doctor wanted to order a medicinal herb that grew only in Italy; or a French chef needed a flavoring from Africa; or a Swiss carpenter wanted special wood from Spain. By writing the binomials (which are mostly Greek or Latin), they could be sure of getting exactly the plant they needed.

Greek and Latin had been used for plant names before, as they were languages understood throughout most of the civilized world. However, many of the names had four or five words, and usage varied from place to place.

Carl Linnaeus was a teenager when he dreamed up a system that would enable the whole world to understand which plant was under discussion. With the Binomial System he invented, each plant was given one name with a capital letter that was also the first name of other similar plants, called the genus. For example, all maples are in the genus, Acer The plant was also given a name that described that particular plant, called the species, which had a lower case letter for the initial, as Acer rubrum, red maple.

Imagine the attitude of Mr. and Mrs. Linnaeus to Carl's announcement that he was going to name every plant in the world this way…

Carl started collecting plants, mounting the types on herbarium sheets and stacking similar types together. As the stacks got higher and fell over, he sorted them into two piles with slightly different characteristics, and gave the latest pile a new genus name.

The major thrust of botany at the time was what is today called Economic Botany, how plants could be used to make money. Using the new system, the herbalist could order Rumex crispus and know he would receive the yellow root of the dock with wavy edged leaves; the chef ordered Glycyrrhiza glabra with full assurance that it would be licorice with smooth leaves that had exactly the right flavor. The carpenter knew that Acer rubrum had reddish wood instead of the pale beige most maples have.

Knowing what the botanists found about each species that set that species apart from the others in that genus helps you know what to look for to tell them apart. The main problem is trying to look into the mind of the botanist who chose the name.

To use this list of words, recognize that many of the words included are root words from Greek and Latin. They may have a hyphen (-) inserted at one end or the other, according to which is most common, but it could be either way. 'Pod-' could be in Lycopodium (wolf foot) or podocarpus (foot seed).

'Mono-' always means 'one'. Some words may have a choice of endings, e.g., 'alb(a,us)' could be written either as 'alba' or 'albus' but either would mean 'white'.

Many of the names will have two or more parts put together. Thus, 'sessiflora' is separated into sessi- and –flora. From this you can figure out that the flowers (flora) do not have stalks (sessile). Chrysanthemum breaks down into chrys- (yellow) and anthemum (blooms), telling you the genus has (or originally had) yellow flowers.

It would take a whole book to list all the epithets and explain why that name was given to this plant. Some say only that this was the name of the botanist who first discovered it, no help except to the botanist's ego, so those are not included here. Some botanists gave the name to honor (or butter up) another botanist.

Others (often ending in –ensis, as carolinensis), tell that this (Carolina) is the place where it was first found, also not included. These are little help but you can't put everything into two words. We are grateful to those botanists who used names (epithets) that help us recognize the species.

Put yourself in the shoes of the botanist who has discovered a new plant. Probably she does not know Latin or Greek, at best has had a single class in it. Out comes a list similar to this one, and she tries to find a word that describes the most important facet that will help her and others to tell this from existing species in the genus.

Remember that the unknown plant has already been keyed to the genus, the capitalized word in the binomial. The species name, with lower case letters, compares the previously unknown plant just to the other species in the genus, not comparing it to all the other plants in the world. If the second word is 'nana', meaning small, it is smaller than the other species in that genus. However, it could be much bigger than many other plants.

Plants are sexy beings, but we understand sexuality better in humans. Pictures of people and their relationships should help you to understand plants a little better. Your time may be spent in figuring out why that picture carries that name. It gives you practice for the same question on the species names. We hope you get a chuckle out of the old-fashioned clip art.

We wish our observant botanist with the newly named plant the best of luck, just as we wish you luck in learning to speak Botanese.

OF ALL THE WONDERFUL THINGS IN THE WONDERFUL UNIVERSE, NOTHING SEEMS TO ME
TO BE MORE SURPRISING
THAN THE PLANTING
OF A SEED
IN THE BLANK EARTH AND THE RESULTS THEREOF.
—JULIE MOIR MESSERVY

They took all the trees and put them in a tree museum,
And they charged all the people a dollar and a half just to see 'em.
Don't it always seem to go That you don't know what you've got
Till it's gone.
They paved paradise and put up a parking lot.
—*Joni Mitchell*

We travel the Milky Way together, trees and men…
trees are travelers in the ordinary sense.
They make journeys, not very extensive ones, it is true;
but our own little comes and goes are only little more than tree-wavings
– many of them not so much.   -- John Muir

LAWN – NATURE UNDER TOTALITARIAN RULE.

The poor ignorant savage even apologized to a tree for having to cut it down
and had sacred groves and woods he kept standing – homes of the gods or his fellow creatures –
whereas his successor, who ungodded nature, ravages the heights
and brings floods, dustbowls and salt pans
into the once fertile lowlands.
Or worse, defoliates to facilitate hunting down his fellow man.
— *Jacob Trapp, The Light of a Thousand Suns*

Gardening requires lots of water – most of it in the form of perspiration.
—Lou Erickson

Gardening is a kind of disease.
It infects you.
You cannot escape it. When you go visiting,
your eyes rove about the garden;
you interrupt the serious cocktail drinking
because of an irresistible impulse to pull a weed.
—Lewis Gannit

On every stem, on every leaf … and at the root of everything that grew,
was a professional specialist in the shape of grub, caterpillar, aphis,
or other expert, whose business it was to devour that   particular part.
–Oliver Wendell Holmes

My garden will never make me famous,
I'm a horticultural ignoramus.
—Ogden Nash

a-:	without, not having, lacking; forget it
ab-:	away from, on the other side
abbreviatus:	shortened version; other guys have more
abietinus:	like a fir tree, short and scratchy
ablocictus:	with a white band , white-girdled, or white-crowned
abortivus:	parts not maturing, not fully developed
abruptus:	becoming shortened, ending quickly, cut off
acalypha:	looks like nettles
acanthocumus:	with spiny hairs or crowns
acaul(e,is):	with no stalk; sorry…
accolus:	lives nearby; saves gas
acephalus:	with no head; old obFred
acer-:	like maple tree or leaf, palmate veining
acerbus:	harsh or sour
acerosus:	needle shaped, sharp and skinny
-aceus:	looks like whatever the first part named is
achilleaefolius:	with feathery leaves like yarrow
acicularis:	needlelike
acidissimus:	very sour and puckery
acidus:	sour tasting
acinaceus:	shaped like a saber or boomerang
acris:	bitter tasting
acrostitchoides:	looking like a swamp fern
acrotriche:	with hairy lips; shave, stupid
actino-:	starlike, a ray, lines out from a center point
aculeatus:	prickly
acuminatus:	sharp tipped, with a weak S curve
acutangulus:	sharply angled, bent back
acutifidus:	sharp cut, like some leaf margins
acutus:	sharp pointed, less than a 90 degree angle
ad-:	toward, close; up front and personal
adenophorus:	gland bearing; like a teenager
adiantoides:	like maidenhair fern, often with a black stem
admirabilis:	noteworthy, attractive; eat your heart out, obMargaret
adnatus:	joined to something or other

adpressu(m,s):	**crowded against, pressing tightly to**
adsurgens:	growing upward, usually curving, ascending
aduncus:	hooked
advenus:	new kid on the block, newly arrived
-aema:	bloody, in blood, red colored
aemulus:	looking like, mimicking, imitating
aerius:	very high, in the air, floating above
aeruginosus:	covered with rusty particles, rust-colored
aesti(vus,valis):	summer, likes hot weather
aethusa:	shiny, glistening

-aeus:	belonging to whatever the first part is
affinis:	very much like something, related
afra:	African, found in Africa
aga-:	much, extremely like something
agavoides:	like a century plant, succulent
ageratifolius:	leaves like floss-flower
aggregat(a,us):	many of something in a herd or cluster
agrarius:	found in fields, domesticated
agrestis:	of fields, either as a crop or weed
agrifolius:	scabby leaved, leaves with scurfy spots
ala(tus)-:	winged, either membranous edges, or flattened sides
alb(a,us):	white, usually flowers but maybe other parts
albescens:	white or becoming whitish with age
albicans:	whitish
albicaulis:	with white stems
albidus:	white
albiflorus:	with white flowers
albifrons:	with white fronds
albispinus:	with white spines
alchemilloides:	like lady's mantle
alcicornis:	with elk-like horns
alectryon:	like a rooster, carrying a crest
aletris:	like cornmeal, dusty, grainy surface
algidus:	cold, likes cold weather, looks frosty
alienus:	foreign, exotic, not native, from elsewhere

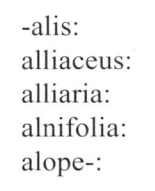

-alis:	having whatever word comes first
alliaceus:	resembling onion or garlic, especially the odor
alliaria:	onion or garlic odor
alnifolia:	with leaves like alder
alope-:	a fox, often said of second rate fruit
alpestris:	nearly alpine, grows in foothills
alpigneus:	of the mountains, grows high up there
alpinus:	found in the mountains
alsine:	bountiful, luxuriant plant
alsodes:	found in or by woodlands, shade loving
altern(us,nans):	with alternating parts, especially leaves
alternifolia:	with alternate leaves
altheoides:	like a mallow
altifrons:	with tall fronds
altissim(a,us):	**very tall**

altus:	tall
alumnus:	strong, flourishing
alyssoides:	like alyssum
amabilis:	lovely, pleasant; eat your heart out, ObMargaret
-amar(a,us):	bitter tasting

amar-:	unfading, long lasting
amaranthoides:	looks like pigweed
amaranticolor:	purplish, especially the stems
amaricaulis:	with a bitter stem
amarus:	bitter
ambar:	having fragrant juice
ambi-:	located around, with a surround of some kind
ambiguus:	doubtful, obscure, hard to key out
ambly-:	blunt ended
ambrosia:	honey, attracts bees and other insects

ambrosioides:	like ragweed
amethys-:	violet colored, usually flowers
ammi-:	umbrella-like plant
ammophilus:	**sand-loving, found in hot sunny locations**
amoinus (amoenus):	pleasing
amorpha:	deformed, shapeless; last blind date
ampel-:	vinelike, clingy, the velcro woman
amphi-:	double, 2 kinds, both sides; AC/DC
amphibius:	grows on land and in water
ampl(i,us)-:	large, bigger than similar species
amplexi-:	clasping, wrapped around something
amygdal-:	almond-like
ana-:	again, repeated
anacanthus:	without spines
anacardioides:	like a cashew plant
ananassa:	pineapple, or resembling pineapple
anceps:	two-headed or edged; back to obMargaret
-anche:	to strangle, twining, possessive
andro-:	male, often refers to stamens
androgynus:	with both sexes
androsaceus:	like rock jasmine
-andrus:	in reference to stamens
anemo-:	of the wind
anethifolius:	with leaves like dill, feathery
aneurus:	without nerves or veins
anfractuosus:	twisted, good old ObFred again
-angion:	vessel, pot
anguinis:	snaky, usually the shape instead of habitat
angularis:	angled, often said of phyllaries
angulatus:	some part bent back sharply
angulosis:	with many corners
angust(atus)-:	narrow, some part in comparison to other species
angustifoli(a,us):	with narrow leaves
anisatum:	anise scented
anisophyllus:	with unequal leaves

anisos:	some part is unequal
annectens:	connecting
annotinus:	one year old
annularis:	with rings
annuus:	lives one year, not perennial
anomalus:	unlike its kind, different from other species
anopetalus:	with upright petals
ante-:	in front or before something
antennaria:	like insect feelers
anthemoides:	like chamomile
-anthemum:	flowered, first part is often a color
anthocrene:	flower-fountain, with flowers erupting
-anthos:	in reference to either flowers or stamens
anthyllidifolius:	with leaves like vetch, pinnate
antiquorum:	**known from ancient times**
antiquus:	known from before history
antirhiniflorus:	with flowers like snapdragon
-anus:	belonging to whatever word came first
aparine:	to cling
apertus:	uncovered, bare
apicul-:	pointed, especially leaf tip
apifera:	bee attracting, sweet smelling
apiifolius:	with leaves like celery
apodus:	with no foot or leaf-stalk, no basal leaves
appendiculatu(m,s):	with a tail
applanatus:	flattened, placed tight against something
applicatus:	joined
apricus:	open, bare, uncovered
apter(atus):	without wings
aquati(cus,ilis):	of water, likes wet places, grows in water
argophyllus:	with silver leaves
arguta:	sharp, toothed
argyro-:	shiny, silvery
argyrocomus:	with silver hairs
aridus:	of dry places, deserts, cliffs
arietinus:	with ram's horns
aril-:	with an envelope, enclosed
-arisa:	having whatever the first part is
aristatus:	pointed, awned, spiky, bearded
aristida:	bristly
aristo-:	best
aristosus:	bearded
armatus:	armed, spiny, ouchy!
armillaris:	with a bracelet or collar
aromaticu(m,s):	odorous, spicy

aqueus:	watery, either habitat or succulent
aquil(egia,inus):	eagle, with sharp claws
aquilegifolius:	leaves like columbine
arachnoides:	spidery, webbed
arborea:	of trees, woody and tall
arboreus:	tree like
arbusculus:	a small tree
arcticus:	**arctic, likes cold places**
arctium:	in reference to a bear
arenari(a,us):	of sand, grows in hot sandy places
arenosis:	of sandy places
areolatus:	pitted, circled
argent(eus,atus):	silvery
argill(a,i)-:	clay or shale color
argillaceus:	grows in clay
arrectus:	raised above, erect
artemisioides:	like wormwood
articulatus:	jointed, with nodes
arundinace(a,us):	reed like
arvensis:	of plowed fields
asarum:	like ginger in flavor or color
ascendens:	climbing
-ascens:	becoming whatever the first part is
asclepiadeus:	like milkweed, especially milky sap
asper:	roughened, scratchy
asprellum:	slightly rough
assimilis:	almost like, similar to
aster:	starry, radiating
astro-:	star-like
-astrum:	looks like but not really, false
ater:	coal black
athyrium:	no scent, odorless
-ati(cus,lis):	place of growth, grows wherever first part says
atomarius:	speckled, dotted
atr(a,i,o)-:	black
atrocarpus:	with dark fruit
atropurpureus:	dark purple
atrorubens:	dark red
atrosanguineus:	dark blood-red
atroviolaceus:	dark violet
atrovirens:	dark green
attenuatus:	**growing to a point**
-atus:	having whatever the first part is
augusti-:	big or wide
augustissimus:	very noticeable

augustus:	majestic
aurantiacus:	golden, orange-red
aurantiacu(m,s):	yellow-orange
aureus:	golden
auriculat(a,us):	with ears sticking out
australis:	southern, y'all
autumnalis:	in the fall season of the year
avicularis:	like a bird, maybe beaked, maybe winged
axillaris:	between stem and branch
azureus:	sky blue

bacca(na,ta,tus):	with berries
balsameus:	like balsam, pleasant piney odor
bambusoides:	like bamboo
baptisia:	to dye, used in coloring cloth
barba:	with a beard; hooked
barbarus:	from a foreign country
barbatus:	some part is barbed or bearded
barbinode:	the plant is bearded at the nodes
basilaris:	**of the bottom**
bellis:	beautiful
bellobatus:	beautiful bramble
benedictus:	blessed, given for medicine
betula:	like birch
bi-:	with two of something
bicolor:	with two colors
biennis:	lives two years
-bilis:	having the ability or capacity
blanda:	mild
blastos:	in reference to a bud
blattari(a,um)-:	a moth, perhaps shape
blepharis:	with a fringe, eyelash
-bolus:	shaped like a ball; to throw
bonus:	good
boreala:	northern, youse guys
botry-:	occurring in a bunch; like grapes; clustered

botrytis: with racemes
bous-; bu: **in reference to oxen; horned; strong**
brachiatus: some part is right-angled
brachy-: some part is short
bracte(atus,osus): the stem has leaflets

brassi-: like cabbage; often having four petals
brevi-: some part is short
brevicaulis: short stemmed
brevipedunculatus: the main flower stem is short
buccinatorius: known, trumpeted forth

bufonius: of toads, with warts
bulb(iferus,osa): bearing bulbs
bulbosus: some part is bulbous
bullatus: blistered, puckery
-bundus: full of; a lot of something

caerule(a,us): deep blue in color
caesia: blue-gray in color, perhaps the stem
caespitos(a,us): growing in tufts or patches; with tufts of hairs
calama-: in reference to a reed
calathinus: some part is basket-like

calcar(atus): wearing spurs, giddy-up cowboy
calcareus: chalk white
calli(s)-: beautiful
callosus: thick skinned, with calluses
calluna: in reference to a broom

caltha: with some part that is cuplike
calsa: something is held high
calv(a,us): bald, with no bristles
caly-: in reference to a cup
came-: dwarf, smaller than usual

campan-: bell shaped
campestris: **of fields, often a farm crop**
campsis: some part is curved
campto-: some part is flexible, not stiff
canaliculatum; something is channeled, with ridges

119

candi(cans,dus):	some part is white, hoary
caneacens:	gray haired, like an old man
canio-:	of dogs, perhaps covered with hair
-canthos:	with thorns
capillaceus:	slender
capillar(e,is):	hair like, in reference to some slender part
capitatus:	in heads, usually referring to flowers
capularis:	with capsules, most likely the fruit
capreolata:	twining, usually the stem is a vine
capsella:	a box, often the fruit
car(eus,inus):	waxy, in reference to a surface
cardi(a,o)-:	of hearts, maybe shape or red color
cardinalis:	red as a cardinal
carduaceas:	thistle like, prickly
carinatua:	keeled like a boat bottom
carn(icus,osus):	fleshy, succulent; or like red meat
carneus:	flesh colored
carp-:	in reference to fruit
carpellum:	seed
caryo-:	nutlike, with nuts for fruit
cassia:	like senna, yellow color or laxative
-cata:	something is located below something
catharticus:	some part is a laxative
cauda-:	some part has tails
caulis:	in reference to the stem
centra:	spurred
centri-:	one hundred; or like a centipede
-cephal(is,um)-:	some part is shaped like a head
-cera:	horned
ceras-:	**like a cherry**
cerates:	horned
chaen:	some part gapes open
chalaz:	look for lumps
chami-:	dwarf, smaller than other species
charis:	graceful
cheilanths:	lip flowered, bilabiate
chelone:	turtle, shaped like turtle head or shell
cheno-:	goose, perhaps graceful neck
chilidon:	swallow (bird), sign of spring
chima-:	winter
chion-:	snow, usually white, may withstand cold
chlamydo-:	cloaked, covered
chloro-:	green
chondrilla:	gum, sticky
-chorea:	in reference to land

chryso-:	golden yellow
cilia(r,t)-:	fringed, with eyelashes
cimic-:	in reference to bugs
cinctus:	**girdled, compressed in the middle**
cinereus :	ash gray
circina(lis,atus):	coiled, with a spiral attachment
cirsium:	with a swollen vein
cissos:	like ivy, appearing like a vine
citr(a,i)-:	like citrus, especially the odor
-cladus:	in reference to a branch
clandestinus:	something is hidden
claus(a,us)-:	closed
clavatus:	club-shaped, like a baseball bat
clavellat(a,is):	with a small club; spurred
clitoria:	shaped like external female organ
clype-:	shaped like a shield
coccifera:	with berries
coccineus:	deep red in color
-coccos:	in reference to a berry
-coccum:	chambers, hollow areas, rooms
cochlea:	spoon shaped
cochleatus:	shaped like a snail shell, coiled
-codon:	shaped like an old fashioned alarm clock bell
coelestinum:	heavenly
collinus:	of hills, perhaps grows on hilly ground
columbinum:	of a dove, with wings
column:	something shaped like a pillar
com(a,o)-:	having hair or a tail like a comet
com-:	together
communis:	grows together, in colonies
commutata:	changeable, varied
comosa:	with a tail, bearded
comosus:	bearded
compactus:	thick, dense
complanatus:	flattened
complexus:	circled, girdled
compressa:	pressed flat, looks ironed
comptus:	adorned, fancied up
con-:	something comes together
concavus:	with a hollow on one side
concolor:	same color all over
condens(atus,us):	crowded together
confertus:	crowded against each other
congestus:	brought together in a cluster
conglomeratus:	many parts crowded together

conifera:	cone bearing, fruits like pine cones
conjugata:	dissimilar parts joined
connata:	inseparable, grown together
connectilis:	joined
cono-:	in reference to cones
consolidus:	parts joined
conspersa:	something is sprinkled, scattered
constrictus:	with a belt pulling the middle inward
contortus:	twisted out of shape
convillaria:	of valleys, growing below the hills
convoluta:	something is rolled up, especially edges
convolv-:	to twine around something
coptis:	cut-leaved
copallinus:	with sticky resin
corallinus:	coral red in color
cordatus:	heart-shaped, especially leaf bases
cordifolius:	heart-shaped leaves
coreo-:	in reference to a bug
coriaceus:	leathery texture, like thick leaves
corn(iculatus,utus):	some part has horns
cornu-:	nodding, like a curved trumpet
coro(lla,na):	in reference to a crown or flower face
corrugatus:	wrinkled like corrugated cardboard boxes
cort(ext,ico):	something about the bark
corydalis:	with a part like the crest of a bird
corymbosum:	flower clusters in corymbs
costatus:	ribbed
costa:	in reference to the midrib of leaf
cotula:	something forms a small cup
cotyle:	flat cap, in reference to the seed leaf
crassi-:	thick
crataegus:	strong
crebrus:	something is close, frequent, repeated
crenatus:	a margin or other part is scalloped around the edge
-crene:	something appears like a fountain
crepis:	in reference to the shape of a boot
crinit(a,us)	having long fringe
crisp(atus,us):	wavy leaved
cristat(a,us):	with a crest
croceus:	dark yellow in color
crota(lon,laria):	noisy, with a rattle
crucifera:	cross formed of leaves or petals
cruentus:	blood red
crusgalli:	cockspur, bearing thorns like rooster legs
crustatus:	crusty

crypto-:	**hidden, often under a hood or cloak**
cteno- :	comb like a grooming tool, or rooster crest
cucull(ata,us):	hooded, some part with a covering
cucullaria:	hoodlike
cucumeri-:	like a cucumber, prickly
cultorum:	domestic crop, of cultivated ground
cuneatus:	wedge shaped, delta
cupri-:	coppery in color
curtipendula:	short hanging parts
curtus:	shortened, abrupt ending
curvatus:	curved shape
curvifolius:	with curved leaves
cuspidata:	with a sharp, rigid point
cyaneus:	deep blue in color, usually flowers
cyan(eus,o)-:	blue color of stem, flower, etc
cyanifolius:	with blue leaves
cyath-:	some part is cup-like
cylindraceus:	something is cylindrical, rod-like
cylindricus:	cylindrical
cymbalaria:	parts are disk-shaped
cymbyformis:	look for something shaped like the bottom of a boat
cyosus:	parts are in clusters
cynodon:	something looks like fangs
cypreus:	coppery colored
-cyst(is,us):	some parts are shaped like bladders or balloons
dactylo(n,oides):	some part looks like fingers
daggea:	dagger
dast-:	a shaggy dog story; hairy
dasya-:	something is thick, like a succulent leaf
dealbatus:	covered with white
debilis:	weak and frail as a fairy princess
dec(a,em)-:	ten of something
deciduus:	dropping leaves in autumn; not evergreen
declinus:	**bent down**
decor(atus,us):	decorative
decumbens:	reclining, creeping
decurrens:	grown together, especially around the stem
deflexa:	turned back
deflexus:	bent sharply down
deformis:	misshapen
delica-:	delicate, tender
deliciosus	delicious, tasty
delphinium:	shaped like a fish or dolphin
deltoides:	triangular
demersus:	prefers underwater habitat

demissus:	low, weak, delicate
dendro-:	in reference to wood or tree, often the size
-dens:	toothed, especially the margin
dens(i,o,us)-:	thick, many close together
dent(a,i,o)-:	tooth
denudatus:	naked, with no covering as a bud with no scales
depauperatum:	starved, impoverished, not flourishing
depressum:	flattened against something
desiccatus:	dried up, but still obvious
desmodium:	a chain
di-:	two of something
dicho(to)-:	forked, splitting in two
dichrous:	two colored
didymus	in pairs, as stamens
difformis:	uneven
digit-:	finger like
dilatatus:	widened, broader than other species
dilectus	valuable, precious
diodia:	**roadside, common**
diplo-:	double
dis-:	apart from
discolor:	blotchy, not evenly colored
disjunctus	separated, not attached
dissectus	lobed, deeply cut
diurnis:	flowering during the day
divaricat(a,us):	widely branching
divaricata:	with varied forms
divergens:	wide spreading
diversi-:	different forms, varying
divisus:	divided, split
dodeca:	twelve of something, many
dolabratus	hatchet-shaped, like a small axe
domesticus	cultivated, grown as a crop
draba:	cress, strong smell
dracontium:	dragon plant
drosera:	**dewy, speckled with shiny spots**
drupaceus:	like a cherry, with a stone heart
dulc(a,i)-:	sweet as sugar
dumos(a,us):	bushy, with many branches
dur(a,i)-:	hard, rough, lasting

e-:	not having, without
ebracteatus:	with no bracts
eburnea:	looking like ivory
echinatus:	bristly, prickly
echium:	in reference to a snake
ecirrhata:	with no tendrils
eclipta:	lacking something other species have
ecornutus:	without any horns like other species have
ecto-:	on the outside
edulus:	edible, tasty
effusu(m,s):	spread out, somewhat floppy
elatine:	a low plant
elatior:	somewhat taller
elegans:	beautiful, elegant
-ella:	smaller than the first part of the word
elo-:	in reference to a marsh
elongatus:	lengthened, stretched out
-elytrum:	in reference to husks
emarginatus:	with a notch at the tip
emeticus:	causes vomiting
endo-:	in reference to the inside of something
ennea-:	nine of some part
ensatus:	**sword-shaped**
-ensim:	comes from the other part of the word
ento-:	within

entomophilus-:	insect loving
epetiolatus:	sessile, without a leaf stalk
epi-:	on, upon, above, among
equ-:	of horses, especially manes and tails
er-:	love, sex, clinging
erect(a,us):	upright
eri(an,o)-:	in reference to something woolly
eri-:	early
ericoides	like heath
-erm:	armed, thorny
erodi(os,um):	heron, beaked
erromenus	strong, robust
erosus:	gnawed, jagged
erubescens	rosy, reddish, blushing
erysimum:	help, especially medicinal
erythro-:	reddish on the other part of the word
-escens:	becoming some other way
esculent(is,um,us):	edible, good to eat
-estris:	growing in the other part of the word
esula:	with bitter juice

eu-:	good, true
eury-:	broad, wide
-eus:	resembles the other part of the word
evectus:	**extended, longer than other species**
evertus:	inside out
ex-:	out of, without
exaltatus:	very tall
excavatus:	some part is hollowed out
excelsior:	taller than other species
excisus:	with something cut away
exigu(a,us):	little, diminutive
eximius:	outstanding, excelling
expandus:	spread out, wider than other species
extensus:	lengthened, longer than other species
extra-:	outside of where it is expected
exudans:	oozing, seeping sap or pitch

fagus:	similar to beech, glossy leaved
falcat(a,us):	something is sickle shaped, like a boomerang
fallax:	false, tricky, perhaps hard to key out
fancicula(ris,tus):	something occurs in bunches
farinaceus:	having starch, perhaps feels like corn meal on surface
farinosus:	powdery, mealy
fasciatus:	**flattened**
fastigiatus:	with branches erect, close
faux:	in the throat; perhaps false, looks like something else
fenestralis:	with windows, often in the leaf
ferox:	ferocious, thorny
ferruginus:	rusty in color
fertilis:	fruitful
fibrillosus:	with fibers
-fid(a,um,us):	some part is divided, cleft, cut
fil(a,i)-:	ferny; threadlike
fimbri-:	some part is fringed
firm(is,us):	strong, firm, solid
fistulosus:	in reference to something hollow, tubular
fissuratus:	fissured, cleft

flabel-:	some part opens like a fan
flaccidus:	soft, droopy
flagell(a,i)-:	look for parts that flail or are whip like
flammeus:	flame colored
flav-:	something is yellow
flex-:	in reference to something that is bent, pliable
floccosus:	with parts that appear woolly
floreplenus:	with double flowers, many petals on the flowers
floribundus:	with many flowers
flor(ida,idus,us):	flowering more than other species
flos-:	in reference to a flower
fluinensis	of a river, grows by riverside
fluitans:	**floating on surface of water**
fluviatilis:	of rivers
foetidus:	strong unpleasant smell
-foli(a,um,us):	in reference to the leaves, foliage
foliolum:	with leaflets
foliosus:	densely leafy
fontinalis:	grows near springs or has flowers that erupt upward
formosus:	beautiful
-fract:	with something that breaks
-fraga:	to break
fragaria:	fragrant
fragrans:	fragrant
frigidus:	grows in cold regions
friseus:	something is pearl gray in color
frondos(a,us):	stems are densely leafy
fructus:	plant produces fruit
frut-:	plant is shrubby, has many stems
-fuga:	to drive away, an insect repellant
fulgida:	some parts are shiny
fulvus:	tawny in color
fumaria:	with a smoky odor
fundi-:	having the shape of a funnel
fungosus:	spongy in texture
funiculus:	some part is like a cord
furcat(a,us):	branches are repeatedly forked
fuscus:	brown, dusty in color or texture
fusiformis.	spindle shaped, wider in the middle

-gaea:	in reference to the earth
gal(acto,ium)-:	milky, especially the sap
-gala:	milk
qaleatus:	helmeted, with a hood over some part
gallicus:	in reference to a rooster, spurred
gemini-: -	some part is paired
gemm(atus,ifera):	in reference to buds
generalis:	a common plant, prevailing
geniculatus:	**some part is bent like a knee**
gentilis:	related
geo-:	of earth or rock, perhaps grows in rocks
geometrizans:	parts appear symmetrical, equal
geophilus:	growing on the ground rather than upright
geran(ium,os):	fruits are like a crane's beak
geron:	old and gray
-geton:	neighbor, a companion plant
-geu(m,s):	in reference to ground, earth
gibba:	humped on one side
-gibbus:	swollen on one side
gigante-:	very large compared to other species
giganthes:	with giant flowers
glabellus:	quite smooth on the surface
glaber:	without hairs
glacialis:	icy, frozen
gladiatus:	sword like in shape
gland(i,u)-:	with obvious glands
glauc(a,us):	with bloom or powder on the surface
glob(o,u)-:	some part has a ball shape
glomer-:	with some parts in clusters
gloriosus:	glorious, superb
glosso-:	some part is tonguelike
gluti-:	sticky
glyceria:	plant has a sweet taste
gnaphalium:	woolly lock of hair
gompho-:	some part looks like a club

-gonus:	sided, first part tells how many sides
gossyp-:	some part is cottony in texture
gracilis:	with slender stems
gramineum:	grasslike in appearance
graminifolia:	with leaves like grass
-gramma:	some parts grow in a line
grandi-:	larger than other species
grandicornis:	large horned
graniti-:	granite, perhaps growing in rocks
granu-:	grainy in texture
gratissimus:	very pleasing
graveolens:	strong smelling
gregarius:	grows in clumps
gris-:	gray in color, perhaps the stem
grossus:	**thick, large**
gummi-:	sticky
guttatus:	speckled in colors
gymnantherus:	naked flowered, perhaps no sepals
gymno-:	nude, hidden, slender
gyno-:	female

haem- :	something, probably flower, is blood red
halim(a,um,us):	growing in salt water
halo-:	salty
ham(a,o)-:	something is hooked
haplo-:	single, other species may have pairs
hast(a,i)-:	forms a triangle with flaring points at base
hebe-:	youthful, downy
hedera-:	viny, ivy like
heli-:	sun, may look like sun or prefer sunny spots
helioscopia:	turning with sun so it always faces the sun
hemera-:	lasting for a day
hemi-:	half
hepat-:	in reference to a liver, lobed
hepaticus:	liver color or shape
hepta-:	seven of something

herbace(a,um,us,i):	green, not woody
hermaphrodit-:	both sexes in one
hesperis:	evening, probably blooms then
hetero-:	of different kinds
hex-:	six of something
hians:	something is open, gaping
hibern(a,alis,um,us):	in winter, perhaps blooms then
hiera:	in reference to a hawk, with a beak, wings?
hieroglyphicus:	appears written on
hilum:	scar, especially on the seed
hippo-:	horse, perhaps coarse hair like a horse tail
hir(s,t)i-:	in reference to anything haired
hirsinus:	with a goat's odor
hirsut-:	hairy
hispid(a,i)-:	hairy, bristly
holo-:	whole, complete
holophyllus:	leaf w/smooth margins
homo-:	same
horizontalis:	crosswise of the main axis
horridul(a,um,us):	extremely prickly
hort(ense,ensis,orum):	of gardens, not found growing wild
-horum:	a mark on the first part of the word
humifus(a,um,us):	grows creeping along the ground
humilifolia:	bears a leaf like hops
humil(e,is):	dwarf, smaller than other species
humistrata:	carpet, perhaps covers the ground
humiusa:	**spreading on ground**
humulus:	bearing soft cones
hyacinth-:	sapphire blue in color
hyalin(a,um,us):	some part is glassy, transparent
hyben(a,um,us):	of winter, perhaps blooms then
hybridus:	hybrid, cross between two species, mongrel?
hybos:	humpbacked
hydro-:	water, probably grows there
hyemalis:	of winter
hygro-:	wet, moist
hymen-:	with a membrane
hyper-:	extreme, beyond, above
hypnoides:	grows like moss
hypo-:	with some part beneath or under another
hypogyn(a,um,us):	ovaries occur under the petals
hypoleuc(a,um,us):	white colored underneath, perhaps the leaves
hystricina:	porcupine-like, with strong prickles
hystrix:	in reference to a hedgehog, spiny

-icola:	lives in the first part of the word
ign(e,i)-:	**fiery in color**
ilicifolius:	with hollylike leaves, perhaps prickly margins
illecebrosus:	prefers to live in the shade; understory plant
ill(i,u)-:	some part is bright, shiny
imber-:	with no beard or spines on the last part of the word
imbrica(cans,tus):	shingled, overlapping like shingles
immaculatis:	spotless
immersus:	grows under water
impatiens:	nervous, bursting, flying apart, shooting seeds
impeditis:	hindered, stopped
imperialis:	kingly, good enough for royalty
implexus:	some part is interwoven
implicatum:	tangled, perhaps seed heads
impressus:	some part appears to be sunken in
incan(a,um,us):	some part is whitish gray
incarnatu(m,s):	look for flesh color
incisi-:	something is cut or slashed
inclinatus:	bent down
incomparabilis:	incomparable, best of the bunch
incomptus:	rude, unadorned, plain, bare
incrassatus:	some part is thickened
indecora:	not pretty, back to ObMargaret
indentatus:	something appears dented
indivisus:	whole, undivided
induratus:	hardened
inermis:	unarmed, has no thorns or prickles
infectorius:	used for dyeing fibers
infestus:	not safe
inflat(a,us):	bladderlike
inflexus:	incurved
infra-:	something is below the other part of the word
infractus:	broken
infundibu-:	funnel shape

inodorus:	without scent
inquinans:	polluting, discolored
inscriptus:	some part looks written on
insertus:	something appears inserted
insignus:	look for something marked
institutus:	grafted, like an unnatural part
intactus:	whole, all one piece
integer:	entire, with a smooth edge
integerrima:	having no teeth
inter-:	among whatever the other part of the word is
interjectus:	put between something
intermedius:	between, found in the middle
interruptus:	stopped, inserted, cut off
intorsus:	turned inward
intortus:	twisted
intra:	thin
intracatus:	involved, tangled
intumescens:	swollen, fatter than other species
intybus:	looks like endive
inundatu(,s):	of floodlands, may germinate or blossom after floods
inversus:	**turned over**
involucratus:	with an involucre, parts below the sepals
io(d,n)anthus:	violet flowered
io-:	violet in color
ipo-:	some part looks like a worm
iris:	rainbow, occurs in many colors
irregularis:	something is uneven
ischaemum:	styptic, used to stop bleeding
-iso(-):	equal
ixo-:	sticky

jubatum:	with a mane, coarse hairs
jugosus:	**with breasts, perhaps grows in mountains**
juncea:	stiff or reed like
juniperi-:	looks like cedar
kali:	alkali, not acid; also listed as large seed

labi(atus,osum,um):	with lips
labilis:	something is slippery
labrusca:	in reference to a wild vine
lacer(a,us):	something appears torn
laciniata:	looks like it was slashed
lact-:	milk, white, perhaps the sap
lacun-:	with pits or holes
lacustris:	of ponds, grows in or near ponds
laeti-:	beautiful
laevigatus:	looking polished
laevis:	smooth
lamina:	in reference to the leaf blade
lan(a,i)-:	some part looks woolly
lanceolata-:	something is shaped like a lance or arrowhead
lappa:	in reference to a bur
lasio-:	downy, woolly, rough
lateri-:	something hangs to one side
lathyrus:	looks like a pea vine
lati-:	some part is wide, broader than other species
lax(i,us)-:	hangs loose and floppy
lei(a,i)-:	a part or parts feel smooth
lentus:	flexible, tough
leon-:	in reference to a lion, with mane or teeth
leopardi-:	spotted as a leopard
lepidium:	with little scales

lepido-:	scaly
lepidus:	**graceful, elegant**
lepis:	with a scale
lepto-:	slender, thin, delicate
leucanthemum:	white flowered
leuco-:	something is white
lign(um,osus)	woody, at least the base, other species may be green
ligul(a,aris,atus):	something is strap shaped
ligustrina:	with leaves like privet
lili-:	lily
limn(i,o)-:	grows in or near a pond
lin(o,um)-:	looks like flax, perhaps with narrow leaves
linear(ea,is):	something is long and narrow or striped
liparis:	with something shining
lirio-:	lyre or harp shape
litho-:	some part is like a stone
littoralis:	of the seashore
litu-:	shaped like a trumpet
-lix:	grows in a spiral pattern
-lob(o,us):	lobed

-lochia:	birth, medical, placenta
loculus:	something hollow, like a chamber
longi-:	a part is long
loph-:	with a crest
lori-:	a part looks like a strap
loti-:	in reference to a lotus
lucid(a,um,ulum,us):	something is shining
lunatus:	shaped like a crescent moon
lupinus:	wolf, destroyer, perhaps poisonous or intrusive
lupu-:	in reference to hops
lute-:	yellow in color
lycium:	prickly
lyco-:	wolf
lygodium:	something is flexible
lyratus:	lyre shaped

macr(a,i,o)-:	large, long
maculat(a,um,osus,us):	some part is spotted
magnus:	larger than the other species
maj(or,us):	larger than the other species
majalis:	in May, probably blooms then
mala(cos,xis):	feminine, weak and politically incorrect
malle-:	some part is like a hammer
mall(a,um,us):	in reference to wool
mam(illa,milar,mosa):	formed like breasts or a nipple
mani-:	of the hand, perhaps like fingers
manicat(a,um,us):	long sleeved, perhaps a covering on some part
mares:	male, husband, married
margari(ali,c,ta):	in reference to pearls
marginat(a,um,us):	referring to the edges
margina(lis,ta,tum,tus):	something is striped or bordered

maritimus:	of the sea, perhaps grows near the sea
marm(a,o)-:	marbled in appearance
marrubium:	containing bitter juice
mas-:	male
matronalis(i,o)-:	in reference to mother or womb
maxillaris:	some part is jaw-like
maximus:	largest of the genus
medicus:	having medicinal properties
medi(a,um,us):	something lies between other parts
medull(a,us,um,aris):	in reference to marrow or pith

mega-:   **some part or the whole plant is large**

melan(a,um,us):	black in color
melli(ssa,tus):	attractive to bees, with honey
melo-:	in reference to melons
membra -:	with thin skin, often transparent
merit(a,um,us):	outstanding
meso-:	in reference to the middle, or mixed with something
meta-:	next to, after
meridianus:	blooming at noonday
mexicanus:	from Mexico
micans:	glittering
micr(a,o):	some part is small
militaris:	military, the shape of a shield
mille-:	many, thousand, often said of delicate leaves
miniat(a,um,us):	colored red
minimus:	smallest of the genus
minor:	smaller than most species
minus:	very small
minut-:	the last part of the word is very small
mirabilis:	remarkable
mirus:	extraordinary
miserrimus:	miserable
mitella:	in reference to a little cap
mitis:	soft, mild
mitr-:	something looks like a four sided hat
mixtus:	mixed
modestus:	modest, as keeping something hidden
-moea:	resembling something
molesta:	troublesome, perhaps prickly or tangled
mollis:	soft, with soft hairs
mollugo:	something is whorled or in a spiral
monilifer(a,um):	with a jeweled necklace
mon(o,a)-:	one of something that others have more
monstrosus:	abnormal
mont(anus,ensis):	of mountains, probably growing there

mor(i,us)-:	resembling a mulberry
-morph(ora):	forms, shaped like the first part of the word
moschat(a,us):	with a musky odor
mucosus:	slimy texture
mucronatum:	with a short point
multi-:	many of something
mund(apulus,us):	neat in appearance
munit(a,um,us):	with ammunition (spines)
muricatus:	rough, with hard points
mus-,musai-,musci-,musco-:	confusing, may refer to mouse, fly,
mosaic or moss	
muscae-:	resembling a fly
muscivoris:	leaves modified for fly eating
muscipul(a,ae):	mouse trap
muscos(a,um,us):	mossy in appearance
muta(bilis,tus):	changeable, especially color
mutilatum:	appearing torn
mutil(a,um,us):	lacking parts that are normal for other species
myo-:	mouse; to shut
myri(a,o)-:	in reference to countless number of some part
myrm-:	associated with ants

nan(a,ae,o,um)-:	dwarf, smaller than the other species
nanellus:	very small
nap(a,i,us)-:	root is turnip shaped
narinos(us,a,um):	with a broad nose
nasturtium:	twisted nose, odor
-nat(a,um,us):	something in identical pairs
natans:	floating on water
nari(ces,x):	in reference to a water snake
nauseos(us,a,um):	vomit, medical property
navicularis:	ship like, boat shaped
nebulosus:	clouded, hidden
neglectus:	grows in waste places, weedy
nelumbifolius:	with leaves like a water lily
nema:	something is as delicate as thread
nemor(alis,osus):	**of woodlands, growing in the shade**
neo-:	new
nepet-:	like catnip
nephro-:	kidney, either in shape or as medicine for
nerv(e,ervis,ia,ius,osus):	in reference to nerves or veins
nesophil(a,um,us):	island loving, grows on shores

-neur(a,on):	the first part of the word has nerves or veins
nictitans:	in reference to the eyelid, blinking, moving
nidiformis:	shaped like a nest
nidus:	a nest
nig(er,ra,ratus,rum):	black in color
nigricans:	swarthy, blackish
nigrofructus:	with black fruit
nipho-:	white as snow or grows in snow
nit(ens,idus,ida,idum):	some part shines
niv(alus,alis,ale):	something is snow white
niveus:	snowy
nivosis:	full of snow
nobilis:	renowned, noble, fit for royalty
noct-:	in reference to night
nocturn(a,um,us):	night blooming
nod(e,is,us):	with a knot or swelling
nodiflorus:	with flowers at nodes
nodosus:	knobby, jointed
nodulosus:	with small nodes
-nodea:	with lumps at the first part of the word
noli-:	beware, perhaps with an irritant or toxin
non-:	not, possibly lacks something the other species have
nonpictus:	not painted
nonpinnatus:	one that is not pinnate
nonscriptus:	one not described or without marks
nosus:	in reference to a nose
-not(a,um,us):	regarding the back or spine
notatus:	some part has a mark
nov(a,ae,um,us):	new
novae angliae:	found in New England
novem-:	nine of something
nox:	in reference to night, perhaps blooms then
nub-:	cloudy, confused
nucifera:	bearing nuts
nud(atus,a,um,us):	naked, bare

nudicaulis:	**with naked stems**
nudifloris:	flowers appear before leaves
num-:	in reference to money, especially coin
nummularifolius:	with leaves like coins
nutans:	some part is nodding
nyct-:	in reference to night
nymph:	loves water

ob-:	just the opposite, backwards
obconicus:	a cone standing on the point
obes(a,um,us):	fat, very fat, like ObFred
obfuscatus:	clouded, confused
oblat(a,um,us):	shaped like Earth or a tangerine
obliquus:	something is slanting, lopsided
obliteratus:	erased
oblong(atus,us):	shaped in a long oval
obolaria:	**looks like a coin**
obovatus:	egg shaped with the wide part away from axis
obscur(a,um,us):	something is hidden
obsoletus:	a part is no longer needed so it is incomplete
obtus(a,um,us)-:	blunt ended
obvallatus:	walled up behind another part
occidentalis:	western in origin, partner
occult(a,um,us):	something is hidden, secret
ocellat(a,um,us):	with small eyes
ochro-:	yellowish in color
oct(a,io) - :	eight of something
octandrus:	with eight stamens
oculatus:	appears to have eyes, with spots
odont-:	tooth shaped
odor(atus,ifera,us):	with a pleasant fragrance
officinal(e,is):	**for sale, especially medicinal**
-oides:	looks like the first part of the word
oleifer(a,us):	oily in texture
olerace(a,um,us):	edible, used to add oil to food
oligo-:	few of something
olitorius:	of vegetable gardens
olivace(a,um,us):	olive green in color
opacus:	not transparent, shaded
operculatus:	with a lid over something
-ophi:	loving something, like a habitat
ophio:	snakelike in appearance
-opho:	carrying something
oppositifolius:	with opposite leaves
-opsi:	looks like the first part of the word
-opter:	in reference to a fern or a wing
-opus:	with a foot
orbiculat(a,um,us):	round as a ball
orchi-:	like an orchid or a testicle
oreo-:	in reference to a mountain, peak, best
orgyalis:	as long as standing with arms extended, about 5 feet
orientalis:	eastern in origin, papa san
ornatus:	decorated, showy

ornith-:   birdlike, perhaps with a wing or beak
orob-:   in reference to vetch
ortho-:   something is straight, upright
-osme: scented, odorous
ossifrag(a,um,us):   resembles an osprey, bird of prey

osteo-:   combining word for bone
ostruthium: like an ostrich, plumed
-osus: full of something
-oton: ear, perhaps the shape
ova(le,lis,tus)   oval in outline

ovi-:   in reference to eggs
ovin(a,us):   like sheep, perhaps woolly
oxy(s)-:   pointed, sharp, acid, sour

pabularius: in reference to pasture or fodder
pachy-:   some part is thicker than other species
palatum: check the throat, say Ahhh
paleaceus: chaffy, having loose broken leaves
paleo-:   of ancient times

pallens: pale in color
palliatus: something is covered, cloak, shroud
palli-:   paler than other species
palmatum: in reference to the hand, fingers; probably veins
palu(dosus,stris):   **of marshs; may grow there**

pan-:   many of something
pandurata: fiddle shaped, with a waistline
paniculata: flower grow in clusters
pannosus: tattered, looking raggedy
papav-:   like a poppy

papillion-:   looks like a butterfly
papyr-:   may be papery, or used to make paper
para-:   near, beside
paradoxus: strange
pard(a,i)-:   as spotted as a leopard

partheno-:   virgin
partit(a,us):   parted in some spot
parv(api,us)-:   smaller than other species
passione: with a thorny crown
patell(a,i)-:   like a kneecap, hollow oval bowl

pat(iens,ula):   spreading
pauci-:   with few of something
pauper-:   poor as a pauper, scarce, few
pauxillus: smaller than others of the species
pavoni-:   peacock

pectin-:	with parts that resemble a comb
ped(a,i)-:	foot, birdfoot or shoe
pedicularis:	in reference to a louse or bugs, perhaps seed shape
pedunc-:	stalked
pellae:	**shadowy, dusky**
peltatus:	shaped like a shield
-pendul(a,us):	to hang or nod
penn(a,i):	like a feather, pinnate
penta-:	five of something
per-:	some extreme part
peramo:	very beautiful
peregrinus:	wandering, immigrant
perennis:	grows for years
perfoliatus:	the stem pokes through leaves
perforatus:	having small 'windows', probably in leaves
persistus:	something hangs on
perspicuus:	some part is transparent
pertusus:	perforated, thrust through
perulatus:	something is like a pocket
petiolatus:	with a stalk on the leaves
petaloides:	another part looks like a petal
petraeus:	rock loving, grows in rocks
phaceli-:	parts form a bundle, clustered
-phagos:	eating, as a carnivorous plant; parts look chewed
phaseol-:	in reference to beans
-pholis:	some part has scales
-phoras:	bearing something; shaped like a basket
photo-:	light as rays, not weight
phyllo-:	in reference to leaves
phyto-:	referring to the whole plants
picro-:	bitter in taste
-pictus:	appears painted
pilea(tus, tum):	**capped, something covered**
pil(eos,osus):	hairy, shaggy
pilosum:	with soft hairs
pingui-:	fat and greasy, like ObFred
pink:	notched like a pinked seam
pinn-:	like a feather, pinnate
piperita:	pepper or peppermint smell or taste
pisiferus(i,o)-:	bearing fruit like peas
placatus:	calm and quiet
plantago:	leaving a footprint, leaf on ground
platanus:	resembles a sycamore
platy-:	some part is broad, flat
pleio-:	few, several

plen(a,i,um):	something is double, full
plicatus:	a part appears plaited, braided, pleated
-pleur:	something hangs at one side
plumosa:	**with a feather, part resembles a plume**
pluri-:	with many of some part
podo-:	like a foot; with a strong flower stalk
-pogon:	with a beard on the other part of the word
-polis:	white in color
polit(a,us):	elegant, polished
ol(us,y)-:	many of something
polylepis:	with many scales
pom(a,e,i):	in reference to an apple
popul-:	like poplar trees
porcinus:	in reference to pigs, pig food
potamophilus:	loving rivers, perhaps growing there
praecox:	very early
pratensis:	of the meadows
pravi-:	some part is crooked
prenans:	something is drooping
princeps:	distinguished
proboscidea:	some part is beaked, with a nose
procerus:	plant is tall
procumbens:	prostrate on the ground
procurrens:	spreading underground
producta:	some part is elongated
projecta:	something is sticking out
prolificus:	**the plant is fruitful**
prostratus:	lying flat
protusa:	protruding, sticking out
pruinosum:	waxy, glistening
prunifolium:	has leaves like a plum tree
pseudo-:	false, pretends to be something else
psycodes:	fragrant, attracts butterflies
ptela:	like an elm seed, winged
pteris:	resembling a fern
ptero-:	winged
pubens:	with soft hairs
pudica:	bashful, modest, hidden
pulch-:	handsome
pulverulenta:	the surface is powdery
pulvinus:	some part is cushioned
pumilus:	the dwarf of the genus
puniceus:	in reference to a pomegranate, red/purple
purgans:	medicinal, laxative
purpure(a, us):	purple in color

141

quadrata:	with four of something
quercifolius**:**	like oak, perhaps lobed leaves
quinatus:	**with five of something**
quinque_:	some part appears in fives

racemos(a,um,us)-:	flowers in pointed cluster, like grapes
radi-:	in reference to the root
radians:	pointing outward
radiatus:	with rays from a center
radicans:	rooting, perhaps where stem touches the ground

radula:	a scraper, with a rough surface
ram(it osus, us):	with branches, twigs
ramosior:	with spokes, of branches
rana-:	in reference to a frog, grows in wetlands
rap(a,ae):	like a turnip, fat roots

raphanus:	radish, appearing quickly
raphe:	in reference to a seam, needle
rar(ipus)-:	scattered, rare
re-:	turned back
reclinatus:	**reclining, bent back**

rect(ior,a,us):	erect, straight, upright
recurvus:	curved back
rediviv(a,um,us):	resusitated, came back to life
redolens:	odorous
reflexus:	bent back

refractus:	something looks broken
reg(alis,ius):	royal, fit for a king
regina(e,rum):	fit for a queen
religiosus:	of churches, used in rituals
remotus:	with parts separated

renbi-:	kidney shaped
repandus:	with a wavy margin
repens:	creeping, bends to the ground
replicatus:	something is folded back
reptans:	creeps like a reptile

resectus:	some part is cut off sharply
resiliens:	recoiling, snaps back
resiniferous:	with pitch or gum
-ret(e,is):	in reference to a network, snare, web
reticulat(a,us):	veined or netted
retro-:	twisted backwards
reversus:	something is reversed, turned back
revoluta:	rolled back, as leaf edge
rhiz(a,o,on,os):	in reference to roots
rhizophyllum:	with rooting leaves
rhodo-:	rosy, like a rose
rhoeas:	like a wild poppy
rhomboideus:	some part is diamond shaped
rhopal(on,ica,icum,icus):	shaped like club or door knocker
rhynch(a,o,um,us)-:	beaked, snout
ribes:	like currants, little fruits
rigens:	something is stiff, rigid
-rim(a,um,us):	to greatest degree, e.g., tallest, whitest
ringens:	something is left gaping open
ripari(a,um,us):	found on river banks
rivularis:	grows by streams
robust(a,um,us):	strong, hard, stout
rorid(a,um,us):	wet, dewy
rosaceus:	like a rose, in rosettes
rose(a,um,us):	rose colored
-rostr(e,is):	some part is beaked
rotata:	shaped like a wheel
rotund-:	round, as leaves or fruit
rub(ella,ellus,ens,us): reddish,rusty	
ruderalis:	grows in rubbish or waste land
rudis:	grows wild
rufus:	rusty, reddish in color
rugosu(m,s):	wrinkled, perhaps the leaves
rumex:	in reference to a dock plant
rup-:	in reference to rocks
rupicola:	growing on cliffs or ledges
rusticans:	in the country
rutido-:	something is rough
rutilans:	reddish, rusty
sabulosus:	of sand, probably grows there
saccat(a,us):	**some part is like a pouch, purse**
sacchar-:	like sugar, sweet
sacculul(i,us):	some part is like a small sack
sacro-:	of churches, perhaps used in rituals

saggit-:	shaped like an arrowhead
sagina:	fattening or with a thick part
salicaria:	like a willow tree or shrub
salix:	with narrow leaves like willows
sals(o,us)-:	salty or perhaps grow in salt water
saltuensis:	of mountain pastures
sanare:	to heal, medicinal properties
sanctus:	holy, sacred, used in rituals
sanguinalis:	styptic, medical properties
sanguine(a,um,us):	blood red in color
sanguinaria:	bloody, either color or styptic
sapidus:	tastes good
sapona(ceus,ria):	soap, perhaps used in cleaning
sapro-:	rotten, feeds on dead matter
sarco-:	fleshy, thick, succulent
sarment:	producing runners
sativus:	cultivated as a crop
sauro-:	in reference to lizards
sax(a,um):	found in or near rocks
saxatilis:	among rocks
scaber:	some part is harsh, rough
scab(rata,er):	scabby, rough
scal-:	like a ladder
scandens:	the plant goes climbing
-scap(a,um,us):	in reference to the flower stalk
scapani-:	like a scapula (shoulder blade), shovel
scariosus:	thin and not green
sceptrum:	looks like a scepter, wand
schisto-:	split
schizo-:	deeply divided

schoeno-:	shoe shaped
scintillans:	**sparkle, glitter**
scler(o)-:	some part is hard
scolopendri(a,um,us):	centipede, perhaps venomous
scop(a,ae):	brush or broom
scopari(a,um,us):	thin branch, broomlike
scopul(i,us):	steep cliff, rock ledge
scorpi-:	some part is like a scorpion curled tail
scrophu-:	in reference to knobs
scut-:	shaped like a shield
scutella:	in reference to a dish
scyph(a,um,us):	some part is cup shaped
sebifer(a,us):	waxy in texture
seclusis:	some part is hidden
sect(a,um,us):	separate, cut off

secundatus:	arranged on one side
segetum:	of grain fields
semi-:	half, partly
semperfloris:	ever-blooming
sempervirens:	the plant is evergreen
senilis:	parts are white haired
sensi(bilis,tivus):	responds quickly, perhaps to touch
-sepal(a,um,us):	combining form for sepal
sepium:	**found along hedges**
sept-:	with something in sevens
sepultis:	some part is buried, hidden
serice(a,um,us):	silky in texture
sericiferus:	silk bearing
serotina:	late in flower or ripening
serpens:	creeping like a serpent
serratifolius:	with saw tooth leaves
ser(a,um,us):	late summer or autumn
sessi-:	leaves with no stalks
setaceus:	with bristles
set(i,o)-:	combining form for bristles
siculiformis:	some part is like a dagger
signatus:	look for something well marked
siliceus:	growing in sand
siliqu(a,um,us):	like a narrow pea pod
silvestrus:	growing in woodlands
simplex:	straight axis, unbranched
sinuatus:	with a wavy margin
-siphon:	in reference to a drinking straw or tube
soboliferus:	with rooting stems
socialis:	growing in colonies

solaris:	**found in sunny locales**
solidus:	dense, solid
somniferus:	sleep producing
spadiceus:	with a spathe or hood
sparsus:	parts are few, far between

spath(a,um,us):	in reference to a double edged sword
speciosus:	showy
spectabilis:	spectacular
speculatus:	shining, as with mirrors
sphaericus:	spherical
spicat(a,um,us):	growing spikes like corn
spiciformis:	spike shaped
spino-:	combining form for spiny
spiralis:	**in a spiral**
splendens:	splendid
spumarius:	something looks like it is frothing, sudsy
spurius:	false, looks like something else
squama:	in reference to scales
squarrosa:	bent-back parts, perhaps phyllaries
-stachy(a,es,o,s):	spiked like wheat heads
stamine(a,um,us):	many or showy stamens
stans:	erect, upright
staphyle(a,um,us):	in bunches like grapes
stella(ria,tus):	some part is like a star
-stemon:	in reference to a stamen or thread
steno-:	some part is narrow
sterilis:	not fertile
-stict(a,os,um,us):	combining form to something spotted
-still(a,ae):	to drip, as from a still
stipata:	crowded
stip(es,ites):	with a stalk, post or trunk
stipitatus:	with a little stalk
stipul(a,ata, atum,atus):	with a bract at base of leaf stalk
stoloni-:	stems travel along ground to form new plants
strat(osa,osum,osus,um):	layered, with a blanket
-strepens:	making rustling sounds
striat(a,um,us):	some part is furrowed, striped
stricta:	straight, erect, upright
stramineus:	straw colored
strepto-:	twisted or in strips
strigosus:	straight haired, lean
strobil-:	in reference to cones
strophe-:	turning
-strot(a,um,us):	covered, spread
struma-:	in reference to tumors or ulcers
st(i,y)l(a,es,is,um):	in reference to a pencil or column
suav(is,eolens):	sweet smelling
sub-:	below or not quite
subcaudatus:	short tailed
suberos(a,um,us):	something like cork

subulat(a,um,us):	in reference to a small pointed tool
succulentus:	fleshy, thick
succumbens:	**sinking down, falling**
suffru-:	shrubby, woody with several stems
suffultus:	supported
sulcat(a,i,um, us):	looks like it was plowed, furrowed
sulcinervius:	with sunken veins
sulphure(a,um,us):	yellow in color or with a sulfur odor
sup(er, ra)-:	something appears above
supinus:	lying flat
surculos(a,um,us):	sending up suckers
suspensus:	some part is hanging
sylv-:	of woodlands
symphoro-:	to bear together
symphytum:	grown together
syn-:	joined together

tabular(e,is):	with a flat top
taenidia:	some part has a narrow strip
tangere:	in reference to touch
tapetodes:	like a tapestry or wall hanging
tardi-:	late, perhaps in blooming
tartarius:	with a rough surface
tazett(a,ae):	shallow cup on a pedestal
tectorum:	roof, shingled
tectus:	hidden, covered
-tegul(a,ae):	roof tiles, top
temulenta:	**drunken**
tenaci-:	clinging, lasting
ten(ellus,er,era):	some part is tender, soft
tenex:	strong, tough, matted
tenebrosis:	found in shady places

tenens:	enduring, perhaps leaves hang on
tenui-:	whole plant or some part is slender
tephrosia:	gray in color
teres:	with a round stem
terminalis:	ending abruptly, with a part at the top
tern(ata,atum,atus,I)-:	with something in threes
terrestris:	grows on land
testa:	with a seed coat
tetra-:	with four of something
tetrahit:	something is four parted
textilis:	used in weaving cloth
thall(a,um,us):	in reference to a twig
thermal(e,is):	from warm spring water
-thri(ch,ches,x):	hairy, bristly
thyoides:	with a pleasant pungent odor
thyrs(a,um,us):	flower cluster is more or less pyramidal
thyrsoides:	like a staff or rod
tigrinus:	striped or spotted
tinctorius:	used in dyeing
titanus:	some part is very large
tomentos(a,um,us):	woolly, with matted hair
tons(a,um,us):	sheared, smooth shaven
torquat(a,um,us):	with a ring or collar, a ring around the collar?
torridus:	grows in very hot places
tort(a,um,us):	crooked, twisted
toxi(carius,icus,iferus):	poisonous
tracheli(a,um,us):	in reference to the trachea, throat, neck
trachy-:	rough texture
trans-:	across
translucens:	some part is translucent
transparens:	transparent
tremul(oides,us).	always moving
tri-:	**three of something**
triangular(e,is):	three sided, as a stem
tricho-:	with hairy something
triginti-:	thirty something
trigon(a,um,us):	triangular
triostemum:	with three bony seeds
tripli-:	three of something identical
tristus:	dull, sad
trivialis:	ordinary
tropis:	in reference to a ship's keel
tropos:	turned
truncatus:	looks cut off at end
tubatus:	trumpet shaped

tuberculatus:	with warts
tuberosa:	with tubers, like potatoes
tub(i,u)-:	tube like
tumidus:	swollen
tunicat(a,um,us):	with a coat, covered
turbi-:	shaped like a toy top
turgid(a,us):	some part is firm, full
tussis :	in reference to a cough, perhaps medicinal
-tyl(a,um,us):	round shield shape, plate
typhina:	like a cattail, velvety
typhinus:	pertaining to fever, perhaps medicinal

uberi(or,us):	dense; fruitful
ud(a,um,us):	from wet places
uliginos(a,um,us):	**in reference to marshes**
ulmifolia:	with leaves like an elm tree
ulmoides:	like the elm
ulmus:	elm, hairy leaves
umbellatus:	umbrella-like
umbilicat(a,um,us):	like a navel in shape, belly button
umbonatus:	with a central projection
umbraculiferus:	bearing umbrellas
umbros(a,um,us):	grows in shade
uncinat(a,um,us):	some part is hooked
unc(ifer,ifera,iferum,us):	something is barbed
undatus:	waved
undulatus:	wavy, as leaf margins
unguis:	with hooks or claws
uni-:	one of something
urban(a,um,us)us:	found in towns, not wild
urceolat(a,um,us):	urn shaped

urens:	burning, stinging
urnigera:	**bearing a pitcher**
uro-:	with a tail on the last part of the word
ursinus:	like a bear, perhaps shaggy or with claws
urtica:	like a nettle, stinging
-ur(a,um,us):	tail of cattle [auroch, taurus]
usitatissimum:	useful in some way
ustula(ta):	burnt, dry
utilis:	useful, perhaps as a tool
utri-:	a bladder, like a balloon
-utus:	having the first part of the word
uvari(a,um,us):	something in a grape-like cluster
uvi-:	with something like a grape hanging
uvulata:	in reference to a palate, hanging

vacilans:	something is swaying
vaccini-:	like blueberries
vagans:	wandering, perhaps floppy or seedy
vagina(lis,tus):	some part is sheathed
validus:	strong, stout
vallicola:	living in a valley
vari-:	different, changeable
varicose:	with irregular swellings
variegat(a,um,us):	variegated colors
varius:	**changeable, diverse**

veget(a,um,us):	with vigorous growth
velutinus:	velvety in texture
venen(atus,osa,osum,osus):	poisonous
venosum:	with many or distinct veins
ventos(a,um,us):	light as or fast as the wind
ventricose:	with a large belly
venustus:	handsome, charming
ver(a,i)-:	true or normal
verbascifoli(a,um,us):	with furry leaves like mullein
verecundus:	some part is bashful, hidden
verm-:	some part looks like a worm
vermiculat(a,um,us):	worm tracks, wrinkled
vern(alis,us):	of springtime
vernico-:	some part looks varnished, shiny
verrucos(a,um,us):	with warty lumps

versicolor:	varied colors
verticillus:	with parts in a whorl or spiral
vesc(a,um,us):	thin and feeble
vesicari(a,um,us):	in reference to a bladder
vespertinus:	of the evening, perhaps blooms then
vestitus:	clothed, covered
vexans:	annoying, irritating, wounding
vexillum:	with a banner
vialis:	of the waysides, lives wild
viatoris:	of the highways
victorialis:	protection against evil
viminalis:	with long slender shoots
vineal(e,is):	of or belonging to vines
vin(i,o)-:	wine, either color or fruit used to make wine
violace(a,um,us):	deep purple color, violet like
vir(ens,idis,ent):	green, verdant
virgatu(m,s):	wand like, twig, striped
virgin(alis, atum):	white; with a membrane
virid(e,i,s):	green in color
viscari(a,um,us):	some part is clammy, sticky
viscidus:	sticky in texture
vit-:	like grape vines
vitellina:	color of egg yolk
vittata-:	some part is striped lengthwise
viviparus:	with asexual reproduction

volubilis:	**twining around things**
vomitorius:	emetic, medicinal
vulga(ris,tus):	ordinary, of the masses, common
vulnerari(a,um,us):	medicine for healing wounds
vuln(era,us):	notched, wounded, marked
vulpina:	foxy, with second-rate frui

W

Whoa! Where'd they Went?

X

xanthinus:	yellow in color
xanth(a,o,um,us))-:	with something yellow
xeros:	growing in dry areas
xylo(n)-:	woody, like a shrub or tree
xylocanthus:	with woody spines
xylorhiz(a,ai):	with a woody root

Y

yuccafolius:	with leaves like yucca
yukonensis:	from the Yukon

Z

zamia:	looking damaged, barren appearance
zantho-:	yellow in color
zebrinus:	zebra striped
zephyros:	west wind, from western hemisphere
zibethinus:	smells like civet cat, strong and yukky
zizania:	weedy; wild rice, wild grain
zona(lis,tus):	zoned, banded
zosterformis:	with a flat stem
zygo-:	joined, a yoke

Arranging a bowl of flowers in the morning
can give a sense of quiet in a busy day –
like writing a poem or saying a prayer.
—Anne Morrow Lindbergh

**The least movement
is of importance to all nature.
The entire ocean
is affected
by a
pebble.
—Blaise Pascal**

Now there is one outstandingly important thing about Spaceship Earth, and that is that
no instruction book came with it.
—Buckminster Fuller

*The best remedy
for those who are afraid, lonely or unhappy
is to go outside, somewhere
where they can be quiet, alone with the heavens, nature and God.
Because only then
does one feel that all is
as it should be
and that God wishes people to be happy among the singular beauty of nature.
—Anne Frank*

**Gardens and flowers have a way
of bringing
people together, drawing them from their homes.
— Clare Ansberry**

*Your first job is to prepare the soil.
The best tool for this is
your neighbor's
garden tiller.
If your neighbor does not own
a garden tiller,
suggest that he buy one.*

—Dave Barry

Bread and butter, devoid of charm
in a drawing room,
is ambrosia
eaten under a tree.

—Elizabeth Von Antrim

Half our life is spent trying to find something to do with the time we have rushed through trying to save.

—Will Rogers

I measure myself
Against a tall tree.
I find that I am much taller,
For I reach right up to the sun
With my eye;
And I reach to the shore of the sea
With my ear.
Nevertheless, I dislike
The way the ants crawl
In and out of my shadow.

—Wallace Stevens

PICK WILDFLOWERS THE SAME WAY YOU DO INTIMATE FRIENDS: SELDOM AND CAREFULLY.

People spend more money on chewing gum than on books, but – after all – you can swap books with anyone.

BIRTH, LIFE AND DEATH. EACH TOOK PLACE ON THE HIDDEN SIDE OF A LEAF.
—TONI MORRISON

Everything in nature is lyrical
in its ideal essence,
tragic in its fate,
And comic
in its existence.
—George Santayana

A garden is never so good as it will be next year.
—Thomas Cooper

Nature
will bear the closest inspection.
She invites us to lay our eye level
with her smallest leaf
and take an insect view
of its plain.
—Henry David Thoreau

PART IV
APPENDICES

If anyone asks, you probably answer that you have never studied botany. It would be hard to survive on earth without knowing botany. Even in the city, plants surround us in landscaping, on the restaurant menu, decorative houseplants, and in clothing such as cotton and linen. We absorb the knowledge painlessly.

You gained knowledge of botany in a series of steps. As a toddler, you picked a dandelion and took it to your mother, the expert on all topics. She told you the name and showed you how to find out if you liked butter by holding it under your chin. That was your **introduction to botany**.

Depending on your family culture, she may have showed you how to make a daisy chain crown, or cut some young leaves for dinner, or made wine, all processes of **applied or economic botany**.

You met **formal botany** in school or scouts or nature oriented programs like summer camp or photography class. The books you used had lovely pictures and little text.

Eventually you end up with huge books bearing scary titles like *Manual of Vascular Plants*, or *Flora of XXX*, or *Horticultural Speciation*. These seldom have pictures at all. You open it up, try to read a description, gulp three times and slam the cover closed.

Like all learning, knowing how the book was written helps. **Plant descriptions** started in the 16th century. The process has been refined, but the order remains much the same. We have arranged the appendices in the same order as the information appears in the books.

First the **whole plant** is described: monocot or dicot (which may be in the beginning pages of the section), growth pattern like tree or grass, and length of life. **Vegetative** parts come next: root, stem, leaf. **Reproductive** parts and processes follow: inflorescense, flower, fruit, seed; the organs first, then corolla. **Flower-fruiting dates**, preferred **habitat(s),** and geographic **range** complete the description.

The breakdown pattern within each of the above is: plant organ, organ parts, common characters, number of parts, color(s), size, kinds of parts.

Leaf description includes shape, margin, tip (apex), base, arrangement, position and orientation. For the sepals (calyx) and petals (corolla), we find number, fusion, orientation, color, shape, size and texture.

You don't need to know all that to tell a dandelion (Taraxacum officinale) from a rose (Rosa sp), but you will when you separate species.

The appendices follow the standard botanical arrangements. The whole plant will be a monocot or a dicot, and have a designated length of life. You have to look hard at a leaf to check out half a dozen characteristics. If there are lots of plants, dig one to find the root type. Does the plant have hairs or armor?

Are male and female organs in the same flower? On the same plant? Are the sepals and petals free or united? What is the name of the pattern of flowers on an inflorescence?

When you first start, you find delight in every new genus you find. Sooner or later, you find you can distinguish species within the genera. Eventually you take pride in finding species in counties where they have not been recorded. You are ready to be called a botanist, even if it is amateur.

PRIMARY DIVISIONS
OF
FLOWERING PLANTS
(ANGIOSPERMS)

	MONOCOTS	DICOTS
# SEED LEAVES	1	2
# FLOWER PETALS	3 or multiples of 3	4 or 5
VEINS	parallel	net
FAMILIES	Grass. Lily, Orchid Iris, Palm, etc	Most tree & shrubs Most forbes

LENGTH OF LIFE

Plants inherit the ability to live for different periods of time. Each pattern has different strengths and weaknesses.

ANNUAL

2007
grow, bloom, seed, die

Annuals are the rednecks of the plant world. They produce many offspring who grow on their own with little care. They are colorful, intrusive and fun.

Most annual seeds will be eaten by bugs, birds and people. Nearly all garden plants are annuals, as well as many of our favorite bedding flowers.

Most "weeds" are annuals, plants that grow in areas that nothing else can stand, pioneer plants that send deep, tenacious roots down through heavy clay or dry sand, loosening soil for "better" plants.

Winter annuals are borderline between annuals and biennials. They grow immediately from summer or autumn seed. They survive winter, sprout the next spring; then bloom, seed and die in spring or summer.

BIENNIAL

2007
grow, may bloom, but usually not

2008
grow, bloom, seed, die

Biennials are sneaky. In the wild, they sprout a small ring of leaves close to the ground with no one noticing them the first year.

If you plant biennial seeds in your flower garden, chances are good that you will hoe them up without realizing they sprouted.

The second year, if they survived, they send up stalks and cover the land with bloom. They scatter a lot of seeds and die.

That means the next year you will probably get no flowers as only basal leaves grow. To get biennial flowers every year, you have to plant for two years in a row. To test this, you might plant yellow one year and pink the next.

PERENNIAL

2007
grow

Delphinium

2008
grow, bloom, seed live 3-7 yrs

Sequoia

3007
first bloom c. 10 years mav live 1000+ vrs

Perennials live for more than two years. You buy plants marked perennial, expecting them to live forever. They may die 3 years later, leaving you feeling abandoned, thinking you killed them. Not necessarily true.

Most perennial garden plants live 10-20 years, but some live only 3 years. All trees and shrubs have to be long-lived perennials to get that kind of size.

As a general rule, you can plan that fast growers and early bloomers will die young.

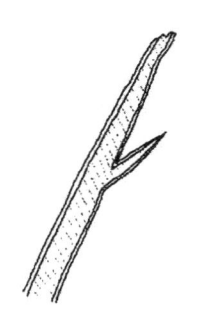

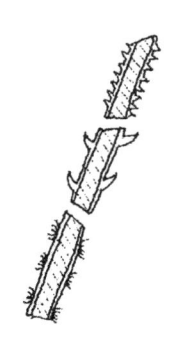

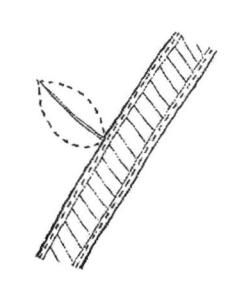

THORN

Grows from wood, may or may not form new branch.

PRICKLE

Extremely varied but comes from skin/bark, not from wood. Rose "thorns" are actually prickles.

SPINE

From leaf, stem or other organ. Adaptation to hot or hostile environment, along with wax coat, water core, etc.

Prickles come in many types with different names. Some names are specialized types, some are versatile: awn, barb, beard, bramble, briar, brier, bristle, bur, comb, cusp, glochid, needle, pin, pricker, rowel, snag, spiculum, spine, spur, thistle, tine

PLANT ARMOR
Any occurrence of thorns, prickles or spines on any part of the plant.
(ARMATURE)

LEAF TIPS

Most leaf tips come to a point.

Acute
Tip is less than a right angle.

Obtuse
Tip is more than a right angle.

Acuminate
Tip narrows abruptly into a weak S curve.

DESCRIBE A LEAF

You have been assigned to find two different leaves and to describe them. "Uh, they are both green" isn't going to cut it. Really look at the differences.

How you describe them when you are starting out:

1. Type	Plain	Is that 1 or 3 leaves?
2. Veins	Like a feather	Like feathers
3. Margin	Small teeth	Uneven
4. Color	Medium green	Green with some red
5. Tip	Pointed	Pointed
6. Base	Uneven	Rounded
7. Stalk	Yes	Yes

If step 1 didn't give you your first clue ("leaves of three, let it be" [actually a compound leaf with three leaflets]), along about step 3 or 4, you hopefully realized that you had a leaf of poison ivy so you didn't pick it. If not, we will see you back here in a couple of weeks…

After a few weeks of botany (without too many in bed with rashes), you will answer the questions like this:

1. Type	Simple	Compound
2. Veins	Pinnate	Pinnate
3. Margins	Serrate	Entire, with notches
4. Color	Medium green	Variable, reddish tinges
5. Tip	Acuminate	Acute
6. Base	Oblique	Tripinnate
7. Stalk	Petiole	Petiole, petiolules
8. Collection	Pressed, mounted	Specimen not collected, toxic

Some leaf tips have long to very long points. Texture is as important to identification as leaf shape. The tip may be a stiff extension of the midrib, or leathery or soft leaf blade material.

STIFF

 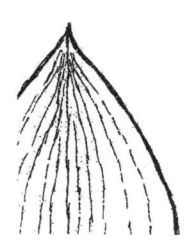 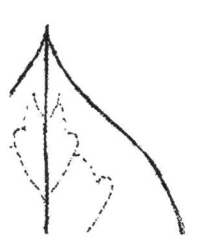

SOFT

Aristate	**Mucronate**	**Cuspidate**	**Cirrhose**	**Caudate**
Long	Rib extension	Rib extension	Very long	Long
Sharp	Sharp	Sharp	Soft curl	Puppy tail
		Leathery		

Some leaf tips are flat or notched in instead of pointed. These are arranged from flat to deep; numbers indicate ratio of notch depth to midpoint of blade.

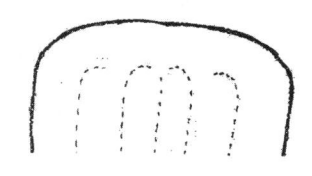   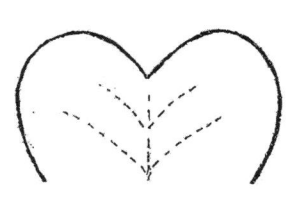

Truncate	**Retuse**	**Emarginate**	**Obcordate**	**Cleft**
Flat	1/16	1/16 to 1/8	1/8 to 1/4	1/4 to 1/2

MORE LEAF TIPS

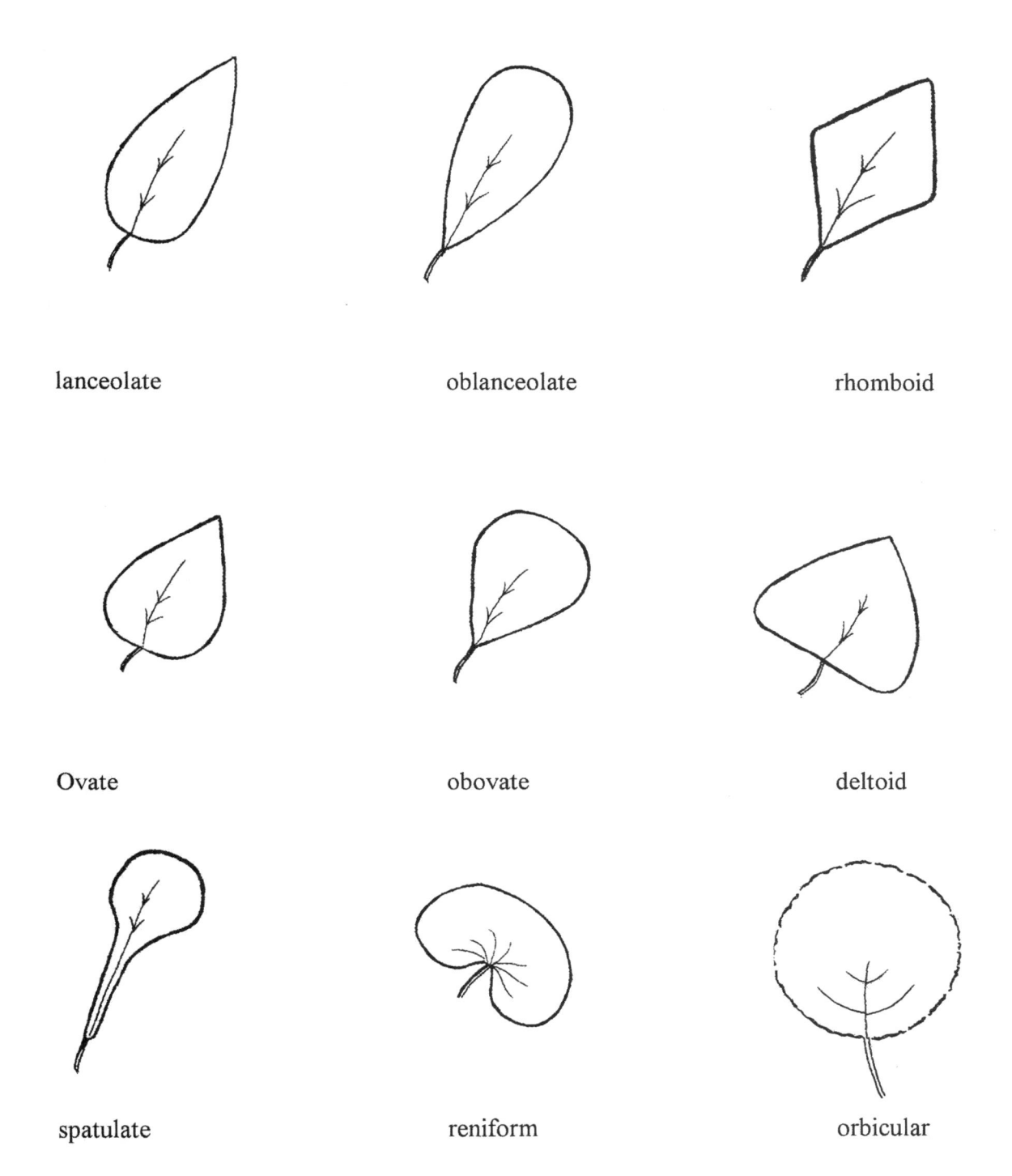

lanceolate

oblanceolate

rhomboid

Ovate

obovate

deltoid

spatulate

reniform

orbicular

SIMPLE LEAF SHAPES

PALMATE

All leaflets *must* start from the same point. There are usually 5, but may range from 3 to about 9, normally an odd number.

PINNATE

Leaflets are arranged on both sides of a central stalk. There may or may not be a center leaflet. If there is, it will have a separate stalk.

The top compound leaf is the most common pinnate (feather) pattern, **imparipinnate.**
It has opposite leaflets and a terminal leaflet, giving an odd number of leaflets.
Center left has an even number of opposite leaves called **paripinnate** (in pairs).
Center right has leaflets arranged alternately called **alternate pinnate.**
Lower right has little "branches" of compound leaflets called **bipinnately compound.**
Lower left has "branches" which have further "branches" of compound leaflets
called **tripinnately compound.**

STRUCTURAL TYPES OF COMPOUND LEAVES

Most leaf bases come to a point

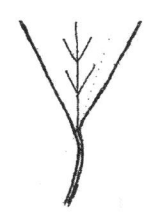 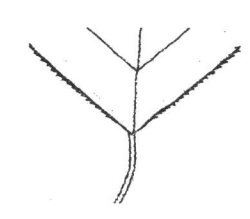 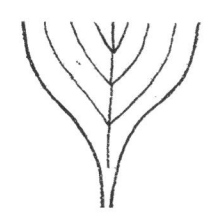

Acute
Base is less than a right angle
(90 degrees)
between left and right margins.

Obtuse

90 to 180 degree angle
between left and right
margins.

Acuminate (Attenuate)

Base narrows abruptly,
then has a weak S curve.

Some leaf bases have two outside lobes that may be sharp or round, with or without a notch around the stalk (petiole). Some sides curve in (concave sinus), other bow out (convex).

Sharp | Round

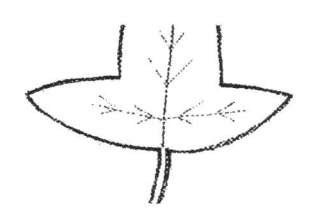 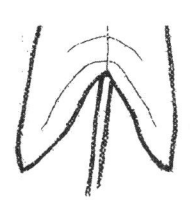 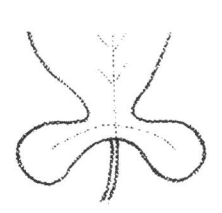 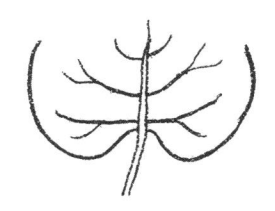

Hastate
Stick out sideways
Flat point/shallow notch
Pointed sinus

Sagittate
Aim down
Notch varies
Little or no sinus

Auriculate
Lobes vary
Notch varies
Concave side

Cordate
Lobes vary
Notch varies
Convex side

Some leaf bases don't have points or notches, may be flat or rounded by the midrib.

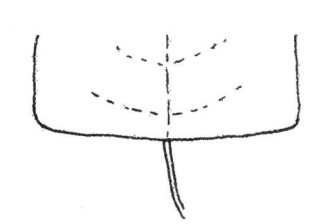 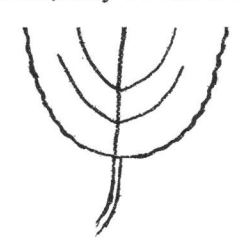 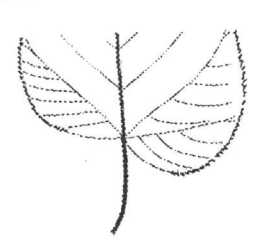

Truncate
Almost flat across midrib

Rounded
Forms arc with margin

Unequal (Oblique)
One side droops

LEAF BASES

The outer edge of leaves, the margin, takes many forms. The work of a leaf consists of collecting sunlight, breathing, preparing and storing food, or storing water. To do all this, they must have vast surface area to collect light and air. Leaves growing near the ground in dense forests may be huge, some nearly a foot long. The same species may produce leaves only an inch or so long in dazzling sunlight. Leaf margins contribute to the variations in order to accomplish their work.

The simplest margin is smooth, but the leaf may be very thick, as in some succulents. The leaf margins may lay flat or raise up in waves or ruffles to increase the amount of available surface.

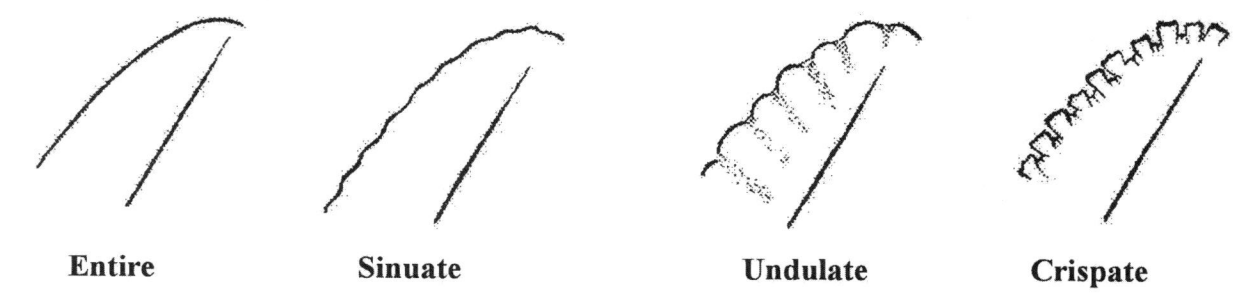

Entire   **Sinuate**   **Undulate**   **Crispate**

Some margins appear to be slashed or nibbled, allowing those edges to wave in the breeze, collecting all available light and air.

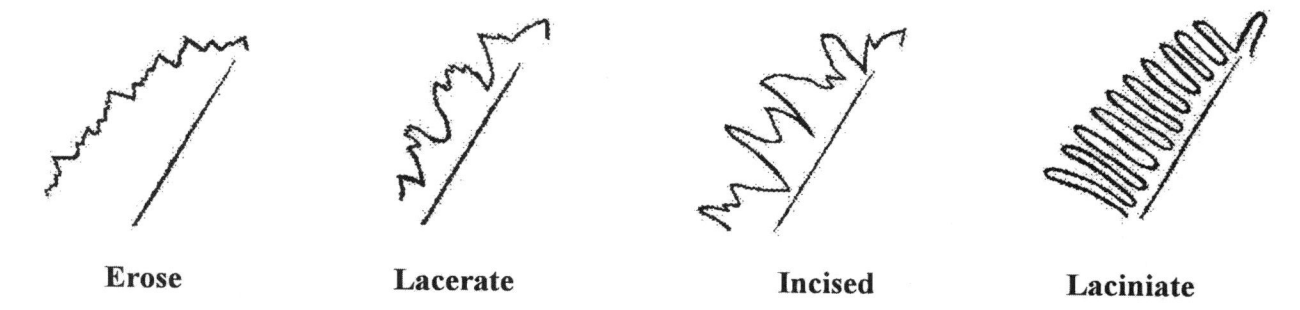

Erose   **Lacerate**   **Incised**   **Laciniate**

Other margins show more neatness, with saw teeth that point forward, large teeth that point outward, or neatly scalloped edges.

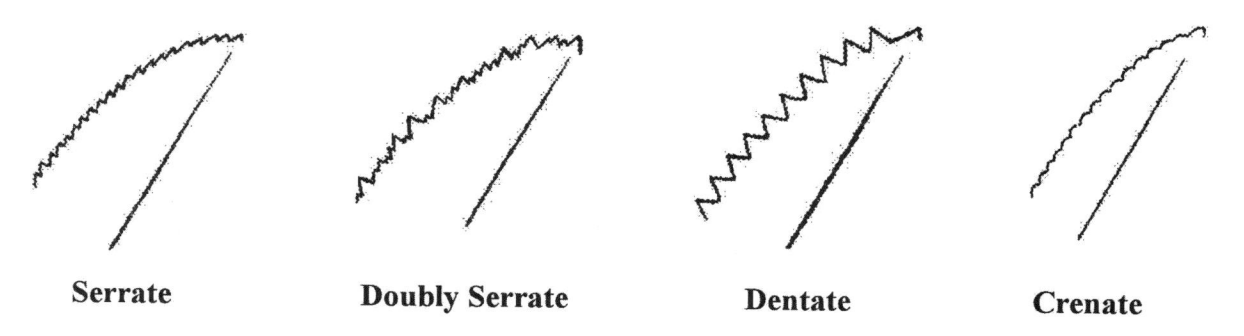

Serrate   **Doubly Serrate**   **Dentate**   **Crenate**

LEAF MARGINS

Most leaves fasten
to the stem with a stalk

Petiolate

Some have no stalk (petiole),
fasten directly to the stem.

Sessile

Some fasten in a circle

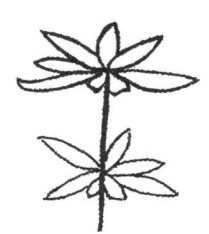

Whorled

Sometimes the petiole
fastens in the middle

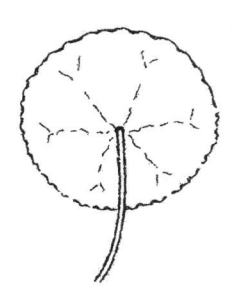

Peltate

Or the stem may pierce
the leaf

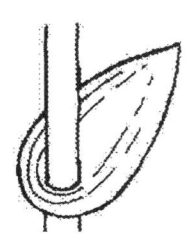

Perfoliate

Or the leaves may join
bases around the stem.

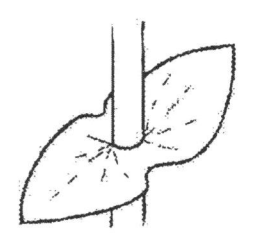

Connate-Perfoliate

The petiole bottom may fasten to the stem in different ways:

Single around
part of the stem

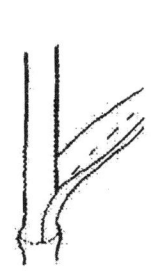

Clasping

Opposite leaves
around each side

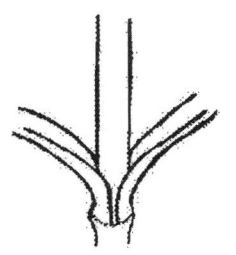

Decurrent

Petiole
surrounding stem

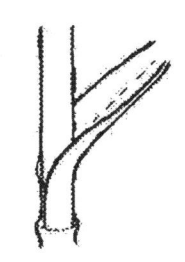

Sheathing

Stipule forms tube
With petiole at bottom

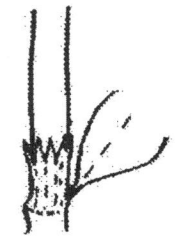

Ocreate

LEAF PLACEMENT AND ATTACHMENT

ROOTS
AND OTHER PARTS
UNDERGROUND
OR ON THE SURFACE

TRUE ROOTS UNDERGROUND

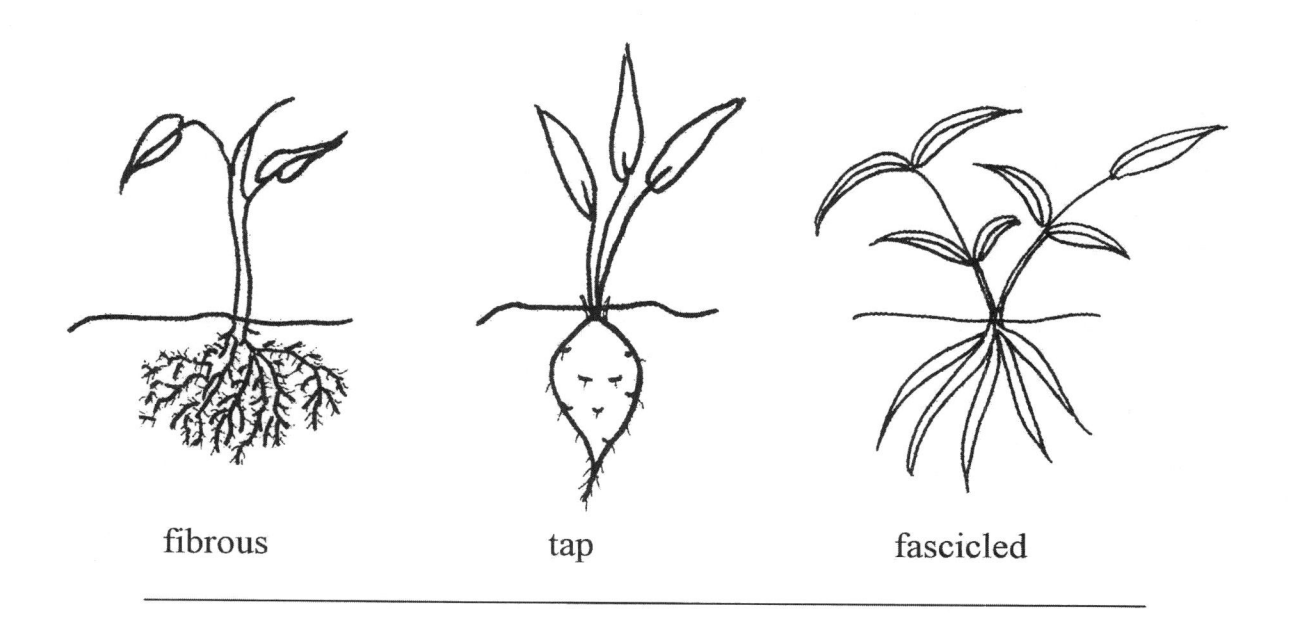

fibrous   tap   fascicled

TRUE ROOTS ABOVEGROUND
(ADVENTITIOUS)

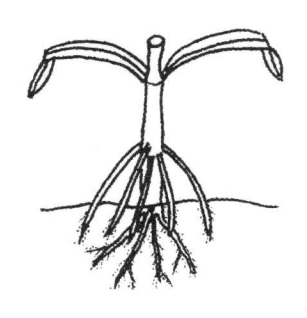

prop roots

aerial roots

OTHER PARTS DOWN AMONG THE ROOTS

STEMS and BRANCHES

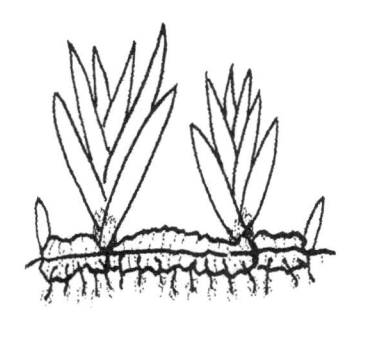 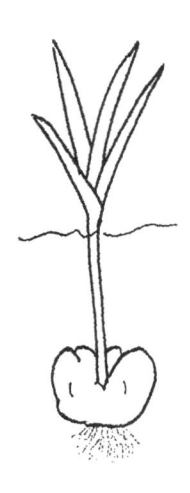 

rhizome   corm   stolon
                               above or below ground

MODIFIED LEAVES
bulbs

 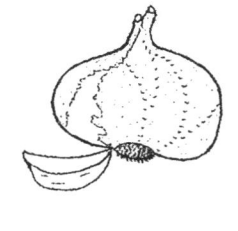

concentric   naked

FRUIT
some legumes

bloom, pollinate
above ground,
ripen underneath

PETALS

If you ask a kid to draw a flower, she will probably draw some form of daisy, what seems like the simplest flower. Actually, a daisy is not simple at all. However, the daisy/aster/sunflower family, Asteraceae (Compositae) is the most common kind of flower, so we are all familiar with it. In Texas, it has been estimated that half of all plants growing there are members of this huge family.

There is tremendous variation within the family, and you may spend hours trying to classify one. In my bloomin' idiot days, I spent a couple of hours daily for 3 days trying to classify a *Psilostrophe tagetina*, woolly paper-flower. A classmate took pity on me and told me to count the petals. They varied from 3 to 7. I had naively assumed that some petals had dropped; they hadn't, they were florets rather than petals. The 6 inch plant with fuzzy leaves did not look like a member of the Aster family, but it was.

The reason "He loves me, he loves me not" works on daisies is because Composites do not have a set number of "petals". When you pull the strap florets from a flower with 10, he doesn't love you, but if the flower has 9 or 11, he does love you.

Take a close look at one of those "petals" that you pulled off – hopefully one where he didn't love you – and you will see that it has notches on the tip. That is an indication that several petals have migrated off to one side of the tube forming a strap, with the notches indicating where the petals united. What you have is not a petal at all, but a complete little strap flower or floret. Anywhere from 3 to a couple of dozen of these surround the outer rim of the daisy/aster/sunflower.

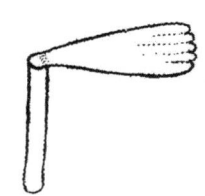

In the center of the head, you find a very different kind of floret. This one is called a disk flower. It is also a tube flower, but the petals cling all the way around the tip of the tube instead of to one side. There are several different patterns that the rudimentary petals may form. Either the strap flower or the disk flower may differ as to the sexual organs present or absent, and these are often hard to see because of their small size in most heads. You will need a strong hand lens to see them at all in most Comps.

Thinking of all flowering plants, the simplest flowers have 3, 4, or 5 petals, or multiples of those numbers, and each petal is separate. Most of the flowers with three petals are Lilies, most of the those with four petals are Crucifers (cabbage family), and nearly all the rest of the families have five loose petals. Their fossils have been found in rock layers that originated from 130 to 60 million years ago.

United petal flowers show the fanciest, and newest, shapes. Fossils from these are in rock from 60 million years ago to the present time. The flowers whose petals are grown together are classed as united petal corollas. Some of the common forms are shown.

INFUNDIBULAR
Funnel shaped flowers with five united petals joined nearly full length with tiny ridges showing the seams.

URCEOLATE
Shaped like a potty, a ball with a rim around the edge. Also called urn shaped.

CAMPANULATE
Bell shaped flowers join five petals with seams
that barely show, but with flaring points

TUBULAR
Any cylindrical united corolla with various
shapes of tips of petals.

ROTATE (back view)
Wheel shaped with a short tube and broad petals
sharply bent back at right angles.

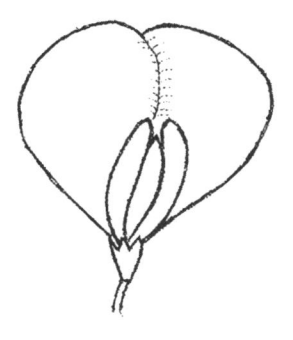

PAPILIONACEOUS
Shaped like a butterfly, with
a back banner, 2 side wings
and a keel.

SALVERFORM
Trumpet shaped with a long slender tube
opening into broad petals, almost at right angles
to the tube

BILABIATE
With 2 lips divided
into unequal sections

169

SEPALS
(CALYX)

Searching through 30+ books, I found none that dealt with variations of the sepals. There probably is a reason, and I may be sorry for attempting it, but since most descriptions include the sepals/calyx, you should know the terms. All flowers are shown from bottom/side view.

FREE SEPALS
(NOT UNITED)

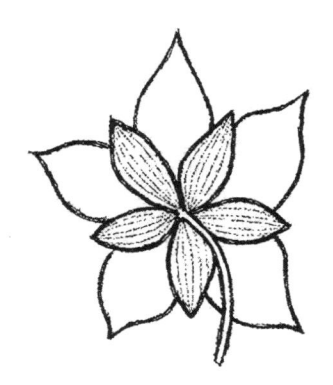

ALTERNATE
(BETWEEN PETALS)
Most common

OPPOSITE
(BEHIND PETALS)

UNITED SEPALS
(CONNATE)

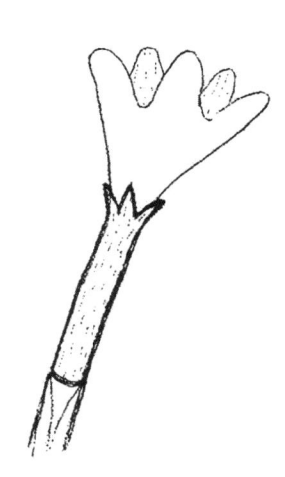

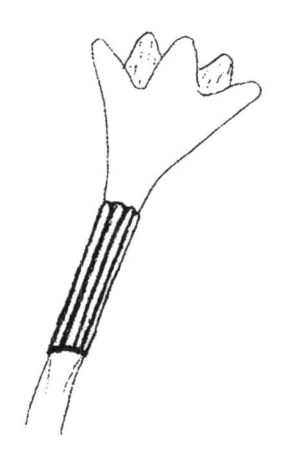

LOBED
(most common)

RIBBED
(one of many variations)

Free petals may appear with either free or united sepals. United petals may have either free or united sepals but more likely to have united.

THINGS THAT LOOK LIKE SEPALS
(BUT AREN'T REALLY)

 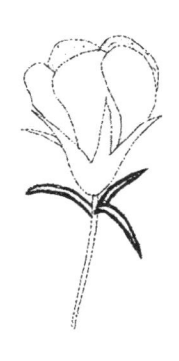 

INVOLUCRE

The involucre lives around the base of flower heads like those found in the aster/daisy/sunflower family, Asteraceae. (Compositae)

The leaflets/bracts are not true sepals so do not form a calyx. We call the individual leaflets or bracts of the involucre "phyllaries."

The mature involucre generally feels sharp or brashy. This protects the ova until fertilization takes place, then holds the seeds as they mature.

INVOLUCEL

The involucel grows around the base of individual flowers which may or may not have a separate involucre around a cluster of flowers, or head.

Involucel leaflets or bracts usually are fewer than the phyllaries of an involucre and often form below (subtend) the true calyx.

Involucel leaflets or bracts probably have a survival purpose, and that would be a good project for you to look up. ☺

EPICALYX

Like the involucel, the epicalyx forms near the true calyx. It also forms in some flowers which may not have a true calyx, like some members of Rosaceae. With other members of Rosaceae, a true calyx is found, but the bracts of the epicalyx become attached (adnate) to the calyx.

If the flower you are trying to identify has extra parts underneath that you can't identify, those parts probably are one of these three.

171

The simplest and most common breeding pattern holds both male and female organs in the same flower. This understandably is called **perfect**. Botanists who feel a need to dazzle you with their brilliance call this **hermaphroditic or monoclinous.**

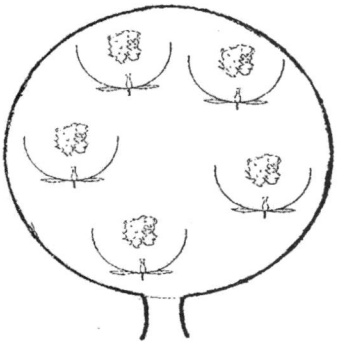

Sometimes the female flowers are on one plant and the male flowers on a separate plant, hopefully not too far apart. Think of this as a sorority house and a fraternity house. The name for this pattern is **dioecious**, with *di-* meaning two, and *ec-* coming from the same root as ecology, meaning house. The title means two houses contain what is needed to make seed.

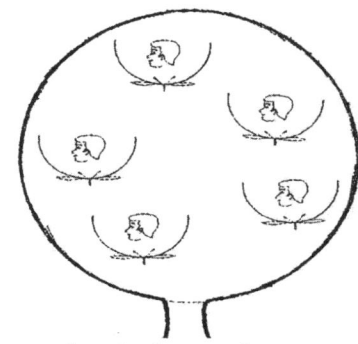

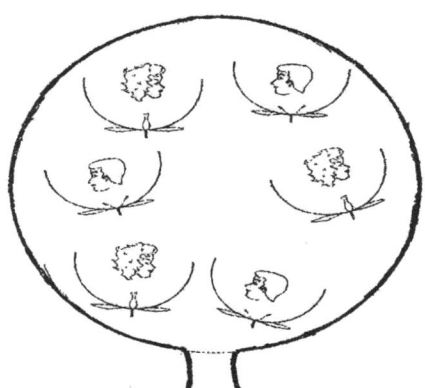

Another breeding pattern is for the plant to produce male flowers and separate female flowers but no flowers containing both sexes. Think of this as a singles bar. This is so popular that it has two names. To compare with dioecious, it is called **monoecious**, *mono-* meaning one, and *ec-* meaning house. This shows that one house contains both sexes. It is also called **diclinous**, *di-* meaning two and *clin-* meaning couch or recliner, indicating that the sexes at least start out in separate beds, er, flowers.

There are other less popular patterns. One is for one plant to have male flowers and the other to have couples -- what might be called a pair tree -- **androdioecious**. Note the similarity to Alaska where all women pair in the dark months.
Rarely, both males and pairs appear on the same plant, as might happen when really cold weather hits. That is called **andromonoecious**. We could be wrong but we think that probably comes from andro- male and mon- moaning.

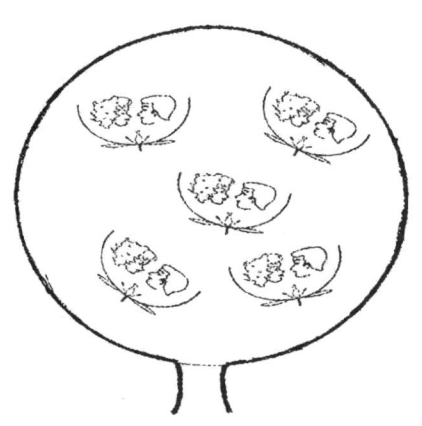

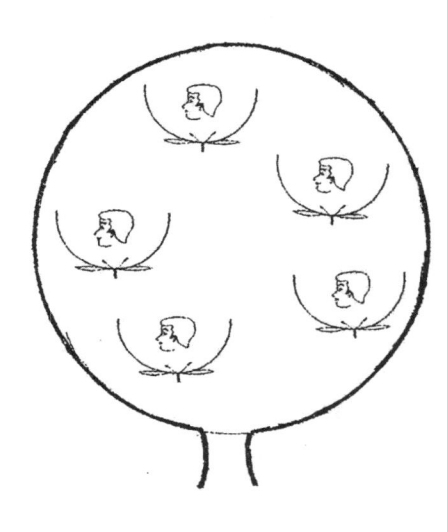

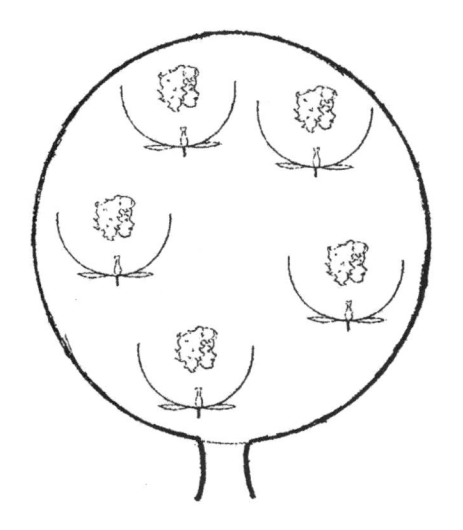

Even less popular is one plant of female flowers and another with couples, **gynodioecious**. Gyno- means female; this is often found at dinner parties and extreme in Washington, D.C., overloaded with secretaries and female pages eagerly seeking partners.
On a single plant, it is **gynomonoecious** and we feel certain the females are moaning one way or another.

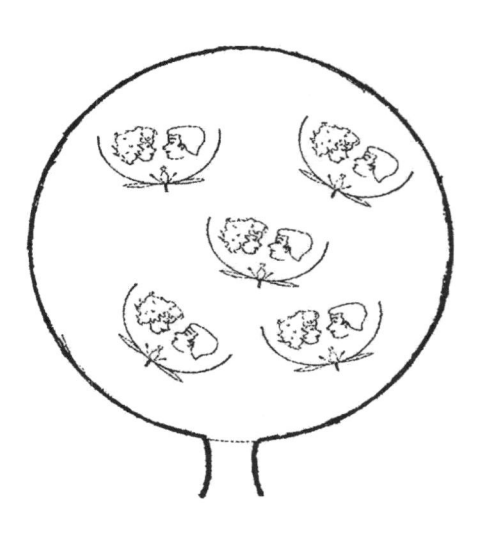

BREEDING PATTERNS

Pollination consists of various ways to transport pollen from the male organ (stamen top - anther) to the female organ (pistil top – stigma), fertilization, to start a new seed. Pollination may be helped by various living critters or environmental conditions.

Pollination depends mostly on insects visiting the flower, called **entomophilous** pollination. Both solitary and social bees collect their biological bribe: honey. Pollen itself provides an excellent source of protein. Beetles top the list of pollen eaters, which also includes moths, flies and wasps. The sweet scent of the flower attracts most insects. However, flies and wasps often lay eggs on rotten meat, so some flowers mimic that disgusting (to us) smell, notably carrion flower, *Smilax herbacea*.

When living creatures spread the pollen, the process is called **zoophily**. Animals who provide the service include insects, birds, and mammals. Some bats sip daintily on nectar, others nibble pollen. Rabbits, bears, deer and other furry mammals retain pollen on their fur and passively deposit it on receptive flowers. These rather rare mammal happenings stay with the general term **zoophily**.

Hummingbirds and a few other birds feed on nectar and pollen and fertilize flowers in the act, called **ornithophily**.

Pollination may be provided by the environment. Wind pollination, **anemophily**, occurs more commonly than most people realize. Trees depend on it more than smaller plants, their increased height allowing wider distribution of the pollen when breezes waft pollen from the stamens, sometimes for long distances, until the pollen settles on a female blossom.

Rarest of all is **hydrophily**, pollination by water. A few water plants, especially on quiet ponds, send up a stalk (scape) with a blossom above water. Pollen sifts unto the surface, is borne away from the male (staminate) bloom and is carried to the female (pistillate) flowers underwater. Other species produce both male organs and female organs underwater, close enough that they can depend on fertilization taking place.

POLLINATION

SEEDS

A seed is the mature result of a flower egg (ovule) being fertilized by pollen.
When you look at a seed such as a bean, *Phaseolus*, you see two parts.

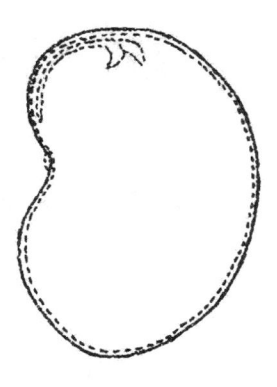

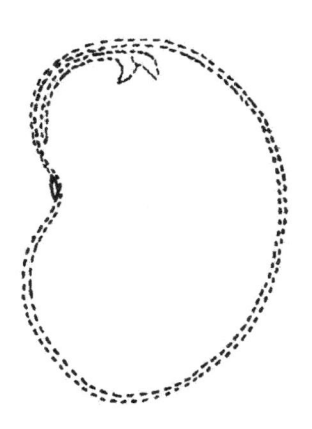

SEED COAT
The protective outer layer

HILUM
The "belly button" where it fastened to
the pod (fruit)

The next time you have a bowl of bean soup, gently pry open one of the beans. Near the middle
you will see a tiny little plant, the **embryo**. The embryo already has true leaves, a stem and a root.

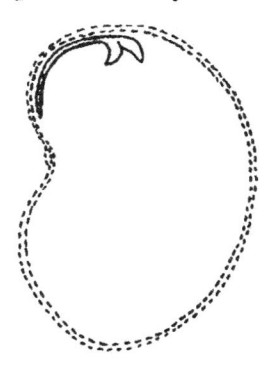

When a seed starts to grow, the seed coat has been softened by moisture.
The first part to break out is the root, **radicle**, part of the embryo.
The radicle will penetrate soil to absorb water and nutrients for the plant to thrive.
The radicle digs in tight to provide a stable
base for the new plant.

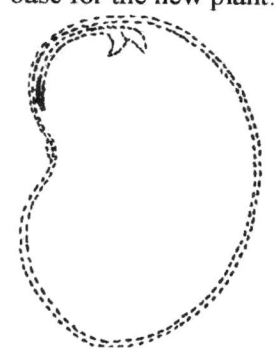

There are two **cotyledons** in a bean seed since beans are dicotyledonous.
Most of the weight of the bean is the seed leaves (cotyledons).
The cotyledons look kind of lumpy and round, not shaped like the real leaves that will appear later.
One of the functions of cotyledons is to provide food storage for the little plant while the radicle learns its job of finding food and water.

Monocotyledonous seeds have only one seed leaf. Corn, *Zea mays*, is an example of a monocot. They generally store their food differently, in **endosperm**, separate from the seed leaf. Order some corn tomorrow for dinner and look those seeds over.

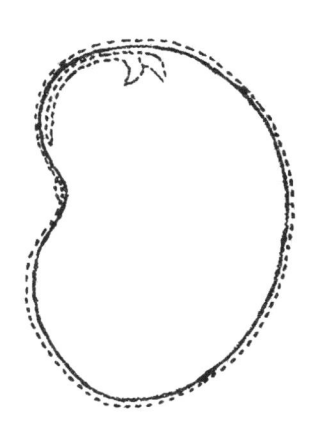

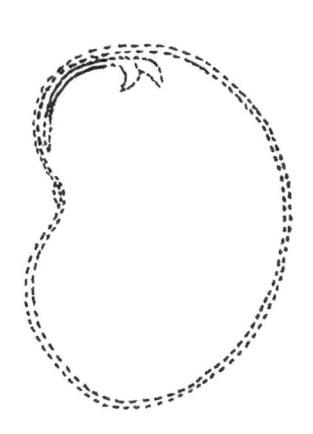

The cotyledons attach to the middle part of the embryo, the **hypocotyl,** which is the beginning of the stem of the new plant.
When the radicle soaks up food and water, those enter the hypocotyl and it grows at an amazing rate.
Humping its back, the hypocotyl presses upward through the soil, searching for sunshine.
When the hypocotyl reaches sunshine, it tugs the cotyledons to the surface and waves them like a banner of victory.
The cotyledons may turn pale green, but they do little work of photosynthesis. The true leaves are waiting in the wings for that job.

The true leaves form the **epicotyl** of the seed.
They look like real leaves rather than like the lumpy cotyledons.
Above ground, they appear between the cotyledons, and the hypocotyl grows on upward to become the true stem of the plant.
The tiny embryo has all the genetic information it needs, to know whether to grow straight upward, or to start producing branches.
In either case, leaves will continue to appear, opposite or alternate, simple or compound, according to genetic instructions.
The seed has done its job of starting a new generation. The plant will carry on as an annual, biennial or perennial.
The plant will flower, be pollinated and produce more seeds for another generation.

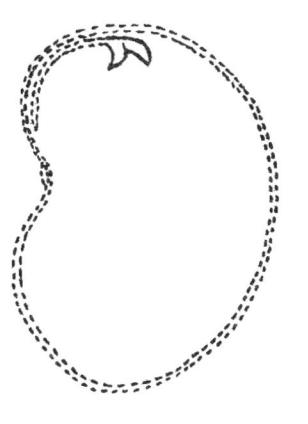

PART V
BIBLIOGRAPHY

Many of the antique books have better descriptions of the basics since they did not have to deal with so much detail as do the present day tomes. To balance this, internet websites have also been included at the end for up to date information.

Bailey, L.H., *Manual of Cultivated Plants most commonly grown in the continental United States and Canada*, MacMillan Publishing Co, New York, 1977.

Bailey, Liberty Hyde and Ethel Zoe Bailey, *Hortus Third*, Cornell University, MacMillan Publishing Co, New York, NY, 1976.

Bergen, Joseph Y., *Elements of Botany, Revised Edition*, Ginn and Company, New York, NY, 1896, 1904.
Bergen, Joseph Y., and Otis W. Caldwell, *Introduction to Botany*, Ginn and Company, New York, 1914.

Britton, Nathaniel Lord, *Manual of the Flora of the Northern States and Canada*, Second Edition, Henry Holt and Company, New York, 1905.

Carpenter, William B, *Vegetable Physiology and Systematic Botany*, Bell and Daldy, Covent Garden, London, 1865.

Coombes, Allen J., *Dictionary of Plant Names*, Timber Press, Portland, OR, 1995.

Correll, Donovan Stewart and Marshall Conring Johnston, *Manual of the Vascular Plants of Texas*, University of Texas, Dallas, TX, 1979.

Dana, Mrs. William Starr, *How to Know the Wild Flowers, a guide to the names, haunts, and habits of our common wild flowers*, Charles Scribner's Sons, New York, 1903.

Forest Service, USDA, *Seeds of Woody Plants in the United States*, Agricultural Handbook # 450, Washington, D.C., 1974, reprint 1989.

Foster, Adriance S. and Ernest M. Gifford, Jr, *Comparative Morphology of Vascular Plants*, W.H. Freeman and Company, San Francisco and London, 1959.

Georgia, Ada E., *A Manual of Weeds, with descriptions of all of the most pernicious and troublesome plants in the United States and Canada, their habits of growth and distribution, with methods of control*, The MacMillan Company, New York, 1937.

Gould, Stephen Jay, *Hen's Teeth and Horse's Toes*, W.W. Norton & Company, New York, London, 1983.

Gray, Asa, LL.D., etc. (!?), *The Botanical Textbook (Sixth Edition), Part I. Structural Botany or Organography on the Basis of Morphology to which is added the Principles of Taxonomy and Phytography, and a Glossary of Botanical Terms* (now that is talkin' lawyer talk as well as botany!), Ivison, Blakeman, Taylor, and Company, New York and Chicago, 1880.

Gray, Asa, *The Elements of Botany for Beginners and for Schools*, American Book Company, New York, Cincinnati, Chicago, 1887.

Gray, Asa, *Gray's School and Field Book of Botany, (Revised Edition)*, American Book Company, New York, NY, 1887.

Gray, Asa, *Manual of the Botany of the Northern United States*, Harvard College, Fifth Edition, Ivison, Blakeman, Taylor, and Company, New York and Chicago, 1868.

Gray, Asa, *Manual of the Botany of the Northern United States*, Harvard College, Sixth Edition, American Book Company, 1889.

Harned, Joseph E., *Wild Flowers of the Alleghanies*, Published by the author, Sincell Printing Company, Oakland, MD, 1931

Radford, Albert E., *Fundamentals of Plant Systematics*, University of North Carolina at Chapel Hill, Harper & Row, New York NY, 1986.

Smith, A. W., *A Gardener's Book of Plant Names*, Harper and Row, New York, 1963.

Strausbaugh, P.D. and Earl L. Core, *Flora of West Virginia* in 5 volumes, West Virginia University Press, Morgantown, West Virginia, 1970.

Swingle, Deane B., *A Textbook of Systematic Botany*, Third Edition, McGraw-Hill Book Company, New York and London, 1946.

Voss, Edward G., *Michigan Flora* in three volumes, Cranbrook Institute of Science; and University of Michigan Herbarium, Ann Arbor, MI, Bulletins #55, 1980; #59, 1984; #61, 1996.

Wood, Alphonso and Oliver R. Willis, *The New American Botanist and Florist, including lessons in the Structure, Life and Growth of Plants*, American Book Company, New York, NY, 1870.

Internet

Learning from the internet is like drinking from a fire hydrant. These sites were helpful in research, but 19 disappeared between research and printing so were deleted. Sometimes typing in the first part of the URL (through .com, .edu, .org, etc) will get you in and you can search for the particular page. Universities and organizations remain more stable than commercial sites; so few commercial sites (xxx.com) are included.

agsites.net. *Links to helpful agricultural botany sites.* http://www.agsites.net/links/botany.html

Agropolis Museum – *Food & Agriculture of the World* – http://www.museum.agropolis.fr/english/index.html

American Forests – *Timeline of American Forests* – http://www.americanforests.org/about_us/history_timeline.php

Answers.com – *Dictionary* – http://www.answers.com/topic/botany http://www.answers.com/topic/stock-car-rail

Agricultural Research Service *Research Timeline* – http://www.ars.usda.gov/is/timeline/1860chron.htm

Baseball-bats.net – *The History of Baseball Bats* – http://www.baseball-bats.net/baseball-bats/baseball-bat-history/index.html

Biology – Online Dictionary – http://www.biology-online.org/dictionary.asp?Term=Obligate

Botanical Society of America – *What is Botany?* – http://www.botany.org/

Botany Online – *Chronology of Significant Historical Developments in the Biological Sciences* – http://www.biologie.uni-hamburg.de/b-online/e01/geschichte.htm

Category-Glossary HCS – http://140.254.84.203/wiki/index.php/Category:Glossary

Common Name Index – http://www.botanical.com/botanical/mgmh/comindxs.html

Dictionary of Botanical Epithets - http://www.winternet.com/~chuckg/dictionary.html

Garden Gate Roots of Botanical Words – http://garden-gate.prairienet.org/botrts.php#top

GardenWeb.com – *Glossary* - http://glossary.gardenweb.com/glossary/

Gardening History Timeline – http://www.gardendigest.com/timegl.htm#start

Garofalo, Michael P., *Spirit of Gardening Timeline* – http://www.gardendigest.com/timegl.htm

Glossaries, Dictionaries and Encyclopedias – Botany – http://www.uni-wuerzburg.de/mineralogie/palbot/glossaries/botany.html

Glossary of Horticulture – http://www.mrgrow.com/content/glossary.htm

Grandiloquent Dictionary – http://www.islandnet.com/~egbird/dict/c.htm

Harvard University, *Library Preservation at Harvard* – http://preserve.harvard.edu/news/preservation/centernamed.html

Horticulture Glossary – http://dodge.unl.edu/HPGlossary/GlossaryA.htm

Huntington Library, Art Collections and Botanical Gardens – *Plant Trivia Botanical Timeline* – http://www.huntington.org/BotanicalDiv/Timeline.html

Integrated Pest Management Resource Centre – *Glossary* - http://www.pestmanagement.co.uk/lib/glossary.shtml

International Journal of Dermatology – Saint Hildegard van Bingen – http://www.dermato.med.br/hds/bibliography/1999saint-hildegard-von-bingen.htm

Iowa State University – e-Library – *Charles E. Bessey* – http://www.lib.iastate.edu/arch/rgrp/13-5-11.html

Kaplan, Donald R., *The Science of Plant Morphology* – American Journal of Botany – http://www.amjbot.org/cgi/content/abstract/88/10/1711

Knotts, Kit & Ben – *Water Gardening Timeline* – http://www.victoria-adventure.org/water_gardening/history/timeline.html

Labyrinth.net – *Garden Timeline* – http://www.labyrinth.net.au/~saul/history/garden.html

Morin, Nancy R. and Richard W. Spellenberg – *Flora of North America – History* – http://www.fna.org/FNA/history.shtml

New York State *Horticultural Study Guide* – http://www.hort.cornell.edu/4hplants/glossary.html

OneLook Dictionary Search (975 dictionaries) – http://www.onelook.com/

Oregon State University – *Biological Pest Controls* – http://oregonstate.edu/~muirp/biocontr.htm

Petersen, Ronald H, *A Guide to Botanical Nomenclature*. A tutorial written to be understood, with links to the Tokyo Code, which is less user friendly. You could buy Van Rheede's Hortus Malabaricus for $875… http//fp.bio.utk.edu/mycology/Nomenclature/nom-intro.htm

Reveal, James L., Dept of Plant Biology, U of Maryland – *History of Systematic Botany* – http://life.umd.edu/emeritus/reveal/pbio/pb250/hist2.html

San Diego Natural History Museum – *Botany* – http://www.sdnhm.org/research/botany/sdweeds.html

Sharp, Gary, *Historical Fishing Events/Development* – http://www.sharpgary.org/FisheryTimeline.html

Saupe, Stephen S., Biology Dept., College of St. Benedict, St. John's University, Collegeville, MN – *The Herbarium and Collecting Techniques*. http://employees.csbsju.edu/SSAUPE/biol308/Lecture/herb_collect.htm

University of Hamburg, Germany (in English), excellent text with illustrations of plant anatomy: http://www.biologie.uni-hamburg.de/b-online/e02/02d.htm

Wilson, Hugh D, PhD, *Field Systematic Botany*. Helpful links. Texas A&M University. http://www.csdl.tamu/FLORA/

Yale University, Medical Historical Library, *The English Physitian* – http://info.med.yale.edu/library/historical/culpeper/culpeper.htm

It is attached to a box, one and a half inches high and less than four inches long, into which it is neatly folded when not in use. The needles are used for dissecting flowers, or other objects, too small to be otherwise handled for analysis. The lenses magnify about **fifteen** *diameters; or, with three lenses, about one-third more.*

A thousand things about forest, field or garden, afford objects of intense interest for daily study.

Prof. ASA GRAY, of Harvard University, our popular American Botanist, says of it: "You are at liberty to call it the "GRAY's MICROSCOPE." I do not think anything better can be made for the money."

Price of Microscope, with two lenses, - $2 00

"   "   " three "   -   2 50

For Sale by

IVISON, BLAKEMAN, TAYLOR & CO.,

138 & 140 Grand St., N. Y., P. O. Box, 1478.

Publishers of Gray's Botanies.

GRAY'S
BOTANIST'S MICROSCOPE.

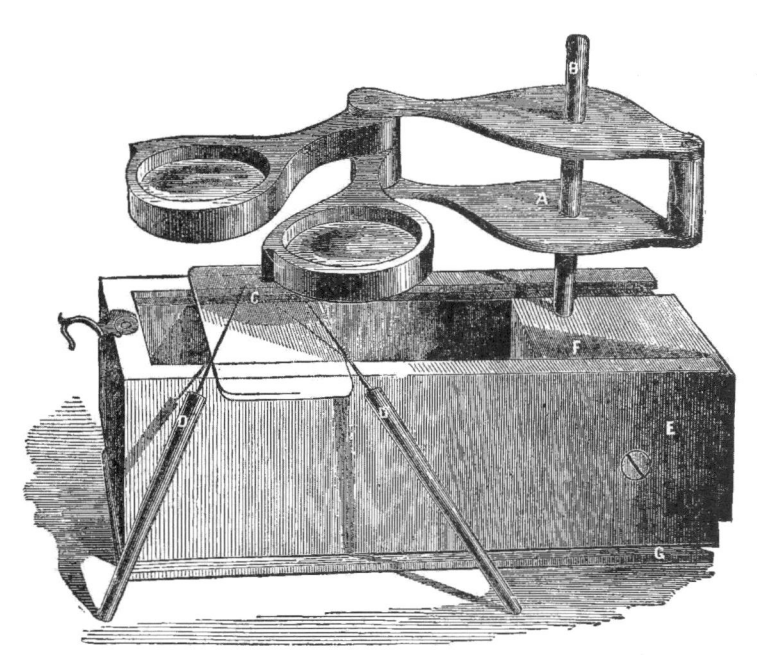

This Convenient Instrument, devised and manu-factured first for the use of the Students in

HARVARD UNIVERSITY,

has given so great satisfaction there, and elsewhere, that we deem it a duty to make it better known, and offer it at a price within the reach of all students.

Barb Short grew up on her grandfather's three-generation Midwest farm, where he shared his love of the land and taught her to read from the Bible and the Sunday funnies. Barb followed him everywhere on the farm, watching him and the uncles prune the flowering shrubs around the stone farmhouse, and graft the apple, pear, peach and cherry trees in the orchard.

They tapped the sugarbush for maple syrup, and grew their own Christmas trees. Barb plucked striped Colorado potato beetles into a tin can of kerosene with her cousins, and held the reins while Belgian horses followed the windrows of hay. They harvested strawberries and raspberries from the family plots, using time-tested growing practices now known as organic. Blackberries, huckleberries, asparagus and many herbs grew wild in the fields and roadsides, free for the picking.

Spring brought dandelions for greens and wine, and bright stalks of rhubarb to pucker winter taste buds. Summer saw the women harvest the acre of garden, and preserve hundreds of quarts of fruits and vegetables in colorful arrays in the basement, with prize-winning county fair projects that often were sent on to 4-H State Show. Fall brought in wheat, navy beans, oats, buckwheat and corn from the fields to stuff the barn and granary for personal use and to be sold for cash. Winter allowed rest with reading and enjoyment of the harvests, with with machinery maintenance and plans for the coming year.

Grandpa had a fifth grade formal education, but never stopped learning. He encouraged Barb to apply for scholarships and go on to college. Equally at home in the library or the field, Barb taught botany to 4-H members in wildflower projects, and at high school and college levels.

Published in a dozen national magazines, including *Better Homes and Gardens, Ranger Rick's Nature Magazine, Michigan Natural Resources Magazine, Farm Journal, Science News* and *Reader's Digest*, she has been quoted as a botanical authority in published articles. Barb has worked for nearly a decade as Editor of GardenWeb.com, with much input on their extensive (accurate but dry – Sorry, Spike!) glossary, and confirming plant patents/registration on Hortiplex.

Barb returned to college when her four children were all in school, and completed 70 hours of graduate work in field biology and business management.

She is a member of American Horticulture Society and Nature Conservancy, and was a charter member of Lady Bird Johnson's National WildflowerCenter in Texas where she attended several courses each year.

She returned to Michigan, where summer often falls on a Tuesday, after a warm decade in Texas where everything sticks or stings, bites or breaks your heart, finding plants in every habitat except tundra.

Botany for Bloomin' Idiots

ORDER FORM   QUANTITY DISCOUNTS AVAILABLE

Postal orders: Barb Short, Fence Row Publishing, 2785 Angle St, Marlette MI 48453

Please send the following books:
I understand that I may return any books for a full refund – for any reason, no questions asked.

___ **Botany for Bloomin' Idiots, Book 1, Talkin' the Talk (2007)**   **$29.95**
___ **Botany for Bloomin' Idiots, Book 2, Walkin' the Walk (2008)**
___ **Botany for Bloomin' Idiots, Book 3, Family Reunion (2009)**

Name
Company
Address
City   **State**   **Zip**   **Telephone**

Shipping   Check or Money Orders only

Priority Mail $5.00 S&H

ORDER FORM   QUANTITY DISCOUNTS AVAILABLE

On-line orders: fencerow@centurytel.net

Postal orders: Barb Short, Fence Row Publishing, 2785 Angle St, Marlette MI 48453

Please send the following books:
I understand that I may return any books for a full refund – for any reason, no questions asked.

___ **Botany for Bloomin' Idiots, Book 1, Talkin' the Talk (2007)**   **$29.95**
___ **Botany for Bloomin' Idiots, Book 2, Walkin' the Walk (2008)**
___ **Botany for Bloomin' Idiots, Book 3, Family Reunion (2009)**

Name
Company
Address
City   **State**   **Zip**   **Telephone**

Sales Tax

Shipping   Check or Money Orders only

Priority Mail $5.00 S&H

ORDER FORM   QUANTITY DISCOUNTS AVAILABLE

On-line orders: fencerow@centurytel.net
Postal orders: Barb Short, Fence Row Publishing, 2785 Angle St, Marlette MI 48453

Please send the following books:
I understand that I may return any books for a full refund – for any reason, no questions asked.

___ **Botany for Bloomin' Idiots, Book 1, Talkin' the Talk (2007)**   **$29.95**
___ **Botany for Bloomin' Idiots, Book 2, Walkin' the Walk (2008)**
___ **Botany for Bloomin' Idiots, Book 3, Family Reunion (2009)**

Name
Company
Address
City   **State**   **Zip**   **Telephone**

Sales Tax

Shipping   Check or Money Orders only

Priority Mail $5.00 S&H